GREEN RUSH

FEVER

By Jimmy Jenkins

GREEN RUSH FEVER IN THE RED HILLS OF NORTH FLORIDA

Observations from the 2019 Florida Agricultural and Mechanical University Industrial Hemp Pilot Project

By Jimmy Jenkins

Foreword

This is a nonfiction book. No characters have been invented and not one event has been fabricated. While all persons and organizations mentioned in this book are actual individuals, established private entities, or public agencies, in some instances their names and identifying characteristics were changed to maintain anonymity. The opinions expressed in this book are those of the author. They do not purport to reflect the opinions or views of the publisher, FAMU|CAFS, UF|IFAS, our hemp industry partner, or any of the employees or staff members of those public or private entities.

The author and the publisher have made every attempt to ensure that the information in this book is correct at the time of print. The author and the publisher do not assume and hereby disclaim any liability to any party for any loss, damage, or disruption caused by errors or omissions, whether such errors or omissions come from negligence, accident, or any other cause. This publication is meant as a source of valuable information for the reader, however it is not meant as a substitute for direct expert assistance. If such assistance is required, the services of a competent professional should be sought. Before farming any industrial hemp, the reader should consult their state department of agriculture and a licensed attorney specializing in matters relating to the cannabis laws in their state.

There have been many changes to the status of industrial hemp law since I began writing this book. The references to state and federal hemp laws that appear in this book were made according to the status of those laws at the time the recounted events occurred. While some industrial hemp laws are discussed herein, this book does not intend to provide an in-depth analysis of those laws, nor does it intend to account for any subsequent changes to state and federal industrial hemp laws that have resulted since the occurrence of the events recounted herein.

Additionally, this book is not intended to serve as an exhaustive manual on hemp farming or cannabis cultivation. Moreover, this material is not intended to represent an exposé or journalistic "hit piece" concerning the state industrial hemp pilot projects as overseen by FAMU|CAFS, UF|IFAS, or FDACS. Instead, this book intends to depict actual events in my life as truthfully as my recollection permits. I have faithfully tried to recreate those events from my memories of them, as assisted by emails, text messages, photographs, and other contemporaneous notes.

SLö Ventures, LLC
2910 Kerry Forest Parkway, #D4-Box 116
Tallahassee, FL 32309-6892

Dr. Althea Henry Jenkins

Dedication

This book is dedicated to my mother, Dr. Althea Henry Jenkins, who has inspired many of the great achievements in my life. My mom graduated from Florida Agricultural and Mechanical University in 1962 with a bachelor's degree in science and began her career as a school librarian for Indian River County in Vero Beach, Florida. She, then, earned her master's in library science from Florida State University in 1972 and became library director at Miami-Dade Community College-Wolfson and Kendall campuses. Althea went on to earn her doctorate in education from Nova Southeastern University in 1977. In 1981, Dr. Jenkins became the director of library services at the University of South Florida, Sarasota/New College campus. While there, she built a new library, moving the New College library holdings from the historic John and Mabel Ringling Mansion on beautiful Sarasota Bay to an updated modern campus library building.

In 1991, Dr. Jenkins was appointed executive director of the Association of College and Research Libraries (ACRL), which at the time was the largest division of the American Library Association (ALA) in Chicago, Illinois. During her decade-long tenure with ACRL, she compiled a list of impressive accomplishments including increasing the financial resources of the association, expanding its publication program and electronic resources, strengthening the continuing education offerings, developing collaborations with higher education and other information-related organizations, and promoting information literacy in student learning.

In 2000, Dr. Jenkins received the Florida State University School of Information Studies Distinguished Alumni award and left the ACRL to join FSU as director of the university libraries. In addition to providing leadership while managing Robert Manning Strozier Library and Dirac Science Library, the two main libraries on FSU's Tallahassee campus, Dr. Jenkins was also responsible for the information resources and services of the other departmental, regional, and international university libraries throughout FSU's campuses and programs.

Dr. Jenkins joyfully retired from work in 2007 and is currently a proud member of both the Bethel Missionary Baptist Church, in Tallahassee, Florida, Pastor Rev. Dr. RB Holmes presiding, and the Tallahassee Alumnae Chapter of Delta Sigma Theta Sorority, Inc., Tallahassee, Florida. Thank you for everything, Mom. Without you, I would not be the person I am today. I love you!

Acknowledgments

This book would not have been possible without the diligence and hard work of the publisher, editors, and graphic designer; all of whom contributed so much to the book's completion and creation. Moreover, I owe an enormous debt of gratitude to those family members and friends who provided detailed and constructive feedback about farming, university research projects, and cannabis cultivation. They freely gave of their time, pushing me to clarify complex concepts and processes.

I also want to acknowledge the administration and staff at FAMU | CAFS for serving as an unrelenting source of positive inspiration throughout the pilot project. This includes, but is not limited to, Dr. Robert W. Taylor, *Dean and Director of Land Grant Programs;* Dr. Charles Weatherford, *Vice President for Research;* Dr. Stephen Leong, *Associate Dean for Research;* Dr. Lambert Kanga, *Professor and Director of FAMU's Center for Biological Control;* and Dr. Alejandro Bolques, *Assistant Professor.* In addition, I would like to thank our hemp industry partners for taking on our small family farm and accepting us as a farming partner. Our hemp industry partners are fine, decent, and hard-working people to whom we wish continued success.

Finally, I am immensely grateful to my wife, Chan, who put up with me throughout the arduous process of farming hemp and writing this book. She gave me her love and constant support through all the late nights and early mornings. Thank you, My Love, for being my muse, my passion, my lover, and best friend.

I Love You!

Table of Contents

Introduction

This book is about my 2019 journey growing more than 2,000 industrial hemp plants on my family's farmlands in the Red Hills Region of Northeast Leon County, Florida. As a farming partner participating in the FAMU|CAFS Industrial Hemp Pilot Project, my family became part of the first group of private farmers in Florida to legally grow and cultivate industrial hemp on their own land. From our humble cottage industry, sugarcane syrup operation to being issued a planting permit in the inaugural year of Florida's industrial hemp pilot project, the following pages and photographs will demonstrate the issues our small family farm encountered along our journey. I will discuss each stage, including the genetics we chose to grow, our soil preparation and cultivation techniques, and how we harvested the plants. I hope you enjoy reading my story.

Part I: Legacy

Legacy (noun) |le-gə-sē| A gift transmitted by or received from an ancestor or predecessor or from the past; a bequest; a gift of property, especially personal property, as money, by will.

Sweet potato storage warehouse at Henry Farms built in the 1950s.

Chapter 1: Our Family Farm

I spent every Thanksgiving, every Christmas, and most summers of my youth on my family's Leon County farm nestled in the Red Hills Region of North Florida. The Red Hills Region is an area of rolling red clay hills that extends for 150 miles along the state borders of Florida, Georgia, and Alabama. The Red Hills section of North Florida is comprised of Leon County and its adjoining counties of Gadsden and Jackson to the west and the counties of Jefferson and Madison to the east.

The North Florida red clay earth, for which the Red Hills region is renowned, is a tightly packed together soil with little or no permeability. The soil's clayey consistency enhances its water storage and nutrient holding qualities and makes the passage of moisture or air difficult. According to geologists and conservationists, for centuries settlers have been attracted to this region's access to fresh water and fertile red soil.

From the indigenous American Indians to the Spaniards lead by explorer Hernando de Soto, the Red Hills region has always been sought as an agriculturally productive homestead. My great-great-grandmother, Ms. Florida Knight moved our family from South Carolina to Florida seeking that same agricultural productivity from her new homestead.

As a woman freed from slavery by the Emancipation Proclamation of 1863, Grandma Knight was able to purchase a sizeable amount of land in northeast Leon County with money she saved through hard work. Over time, Grandma Knight parceled out the land to her three adult sons: George Henry Sr., Allen Henry Sr., and John Anderson. My grandfather Sam Sr., born in 1912, is a direct descendant of George Henry Sr. "Granddaddy Sam," as I affectionately called him, and his wife our grandmother Florence, had fourteen children together. My mother, Althea, is the third eldest of the fourteen. Those children along with the other descendants of Grandma Knight have continuously owned this land and farmed portions of it in maintenance of our family legacy.

I remember helping Granddaddy Sam harvest his crop of watermelons while visiting the family farm during the summer when I was around ten years old. He drove his old green Ford truck slowly along the turn row next to the field while his sons my uncles Timothy and Willie Sr. walked alongside picking watermelons from the rows and carrying them to the truck. How was it that the sun beaming on that North Florida red clay field seemed so much hotter than it ever had further south in my native hometown of Gifford, Florida? I dared not carry any melon larger than a medium-sized one, for fear of dropping it while lumbering across the red clay field.

I marveled at how my uncles effortlessly strode across the rows with their watermelons held high above their shoulders in each hand. After the melons were loaded, we all jumped onto the truck and rode into town, where Granddaddy Sam sold them in local neighborhoods like Frenchtown, the Bond Community, and Griffin Heights. After he had made a few sales, Granddaddy Sam rewarded my meager work efforts with an ice-cold bottle of soda.

Despite the nostalgic feelings that emanated from this wonderful family memory, I had no desire to farm my parcel of family land in modern times. Don't get me wrong. I am proud of my family's farming background. In fact, I earnestly believe that the land, the crops we harvest, and the communal spirit with which we complete that work are all a part of our family legacy. However, I remembered the rigors of farming work from my youth. I had not undertaken any activity nearly that strenuous since moving to Leon County from New York City almost two decades earlier to help care for my ill father, J.C. I sought no greater involvement in the farming than consumption of the harvest.

After my father passed away from his ailments, I did not return to my former role as a criminal prosecutor. Nor did I seek refuge from grieving his loss by plunging myself into a back-to-nature farming scenario on my plot of family farmland. Instead, I opted for a less-stressful lawyering position with a local publishing company in order to stay closer to my aging mother. She had subsequently retired to her residence on the family property in Leon County.

Historically, my family had found success growing sugarcane, sweet potatoes, watermelons, and other traditional row crops on portions of the more than 100-plus acres of our family-owned land. In fact, we still used our heirloom sugarcane press and large metal kettle to make sugarcane syrup. Several family members firmly believed that the longevity of our homecooked cottage industry syrup operations foretold of possible commercial shelf success in today's market. As a result of these strong evocations by family members, my wife and I decided to research the family sugarcane syrup operations as a potential business venture.

It was this initial journey that led to our participation in the inaugural year of the state of Florida's industrial hemp pilot project. As this story will show, I just happened to be in the right place, at the right time with an outlook that was ready to ride the Green Rush. The term Green Rush describes the emerging billion-dollar agribusiness industry involving cannabis in the U.S. and Canada, including the establishment of cannabis-related enterprises, the movement of cannabis within the borders, and the sale of cannabis-related products and stocks.

I wasn't trying to achieve an historic accomplishment, nor was I originally tempted by the allure of acquiring a foothold in a Green Rush industry. I say "the right place" ardently because that place turned out to be the farmland that has served as the cornerstone for my family's legacy. I say a mindset "ready to ride the Green Rush" because our initial interest involved farming sugarcane which was different from farming in the emerging cannabis agribusiness industry.

Chapter 2: The Heirlooms

Making sugarcane syrup is a southern tradition that dates back to the days of Florida's early settlers. In fact, sugarcane syrup was the primary sweetener for small southern farming communities where refined sugar was difficult or impossible to obtain. Sugarcane syrup is a dark, honey-colored substance, with a slightly sharp taste, that is essentially made from the juice of the sugarcane. Many people prefer the naturally sweet sugarcane syrup to the over-refined sweetness of high fructose corn syrup or white sugar products.

Our family's sugarcane harvest always involved a syrup boiling event or "cane grinding," as we call it. The cane grinding has always been a happy time meant for community gathering, typically occurring between Thanksgiving and the

New Year. Music was played, fish was fried, and libations were poured in celebration of another bountiful harvest. We invited close family friends, who accepted with much aplomb, to join the family in grinding the harvested sugarcane. Folks always arrived excited to see the family's heirloom sugarcane press and kettle put back into service.

The operation of the cane press involved crushing the cane stalk to extract its juice. Our press is a horizontal tractor powered press. The farmers who had harvested their crops produced truckloads of freshly cut cane and piled it next to the press mill. Loads of cut cane stalks were fed into the turning rollers of the press, where the outer leaves on the stalks were stripped by the crushing and squeezing of the press rollers.

The Henry Family heirloom sugarcane press equipment.

We never made a distinction between the various varieties of sugarcane that were piled up for pressing. It was all pressed together, which, in hindsight, made the flavor unique but madly inconsistent in texture and viscosity. The juice extracted by the press was funneled to a holding tank fitted with a filter to remove any coarse cane stalk fibers before the juice entered the tank. The extracted juice was then channeled to the 100-gallon, double-ring, cast-iron kettle for continuous boiling at approximately 215°F for three or more hours.

*The one hundred gallon, double-ring, cast-iron kettle in
which extracted sugarcane juice is boiled.*

At this temperature, impurities in the sugarcane juice separate from the liquid and float to the top as a gray film that is removed with long-handled skimmers. What remains is a syrup safe for human consumption. We consistently tested the temperature of the boiling syrup and checked its thickness to determine the proper time to perform the "strike," or extinguish the fire from the hearth and remove the syrup from the furnace.

When the strike was called, we brought out bucket-shaped dippers to scoop up and collect the thick syrup. Folks eagerly waited in an assembly line at a nearby bottling station with clean towels, glass bottles, caps, printed labels, and boxes. As efficient as our cottage industry process appeared, the inconsistent texture and flavor of each syrup strike forestalled any attempt to scale the operation for commercial market distribution. That inconsistency in the final product would increase the costs for bonding, insurance, and any other financial security required to sell items from grocery store shelves.

The interested family members discussed available options, but still revealed an air of hesitancy. When I asked them about their tentativeness, they replied that we would need to grow much more sugarcane if there was going to be enough syrup produced to even require any financial securities. I was momentarily stymied by this revelation. Until then, it had not occurred to me that, despite all I had discovered about our syrup operation, growing the sugarcane would be the most necessary part. They essentially told me that we had to grow enough of the crop before we would have enough of it to decide what to do.

This straightforward sentiment would later ring true throughout our hemp grow, from the first planting of our crop to its eventual harvest. The family syrup cookers pointed me in the direction of the Florida Agricultural and

Mechanical University, College of Agriculture and Food Sciences (FAMU│CAFS) research farm to get information about growing sugarcane. It seemed that if I was going to commit to the syrup making venture, then I would need to prepare for farming sugarcane before the next planting season began.

Despite discovering so much about the family syrup making operation, I knew nothing about modern farming activities. My continued interest in this venture required finding the most farm knowledgeable person I knew to answer these questions. That would be my uncle, Darrell Sr.

Chapter 3: The Sugarcane Farming Plan

Uncle Darrell Sr. grew up in Central Florida on his family's fruit orchard. He had become one of the favorite sons-in-law of the family upon marrying my aunt, Vicky, Granddaddy Sam and Grandma Florence's youngest daughter. A graduate of FAMU|CAFS, my uncle had worked with agriculture and farming at the federal, state, and local levels ever since I was in high school. He worked for the U.S. Department of Agriculture (USDA) in various capacities for more than three decades. Even in his retirement he remained extremely knowledgeable about the status of the agriculture industry in Florida.

My uncle explained that the size of Florida results in the state having three different growing regions. Those regions are the northern, central, and southern regions. The boundaries of these regions are recognized geographical landmarks. Our family farm in the Red Hills naturally resides in the northern region.

My uncle further explained that the state has three distinct growing seasons: spring, summer, and winter. He emphasized that the timing of these seasons will vary in each of the different growing regions because each region has a slightly different climate than the other. The most impactful difference is the possibility of freezing temperatures in the northern and central regions. It was important to know these factors about the natural resources on which the family farm would rely. I had grown up hearing about the northern and central Florida citrus freezes of the 1890s and 1980s. However, I had not considered the full effect of those events from the aspect of a produce grower before.

The FAMU|CAFS information noted that sugarcane is a giant, robust, tropical perennial grass native to Asia where it has been grown for over 4,000 years. It can have up to a 15-month growing cycle: annual planting typically occurs between September and January of one year with harvest occurring between late September and late April of the following year. However, sugarcane had been successfully planted as early as March at some North Florida farms.

Farming sugarcane did not seem impossible to undertake, as explained by the university literature. The one fact that gave me pause was the annual planting aspect of the crop. With only one season to grow everything needed for manufacturing until the next year, there was little room for farming error. I no longer wondered why the syrup cookers of my family had said we had to grow enough of the crop before we would have enough of it to decide what to do. As I learned more about growing sugarcane, I also took note of the cautionary pitfalls Uncle Darrell Sr. warned had caused the loss of many small family-owned farms.

First, public policies that advocated large-scale farming put enormous pressure on all small family farms, of any race or ethnicity, to become larger just to compete. This large-scale, industrial agriculture business model was first developed in the 1970s under USDA Secretary Earl "Rusty" Butz. At that time, a number of agricultural economists criticized Rusty Butz's "go big or go home" policies as causing the biggest rural crisis since the Depression of the 1930s.

The result of those policies being a significant decline in the number of all farms in the United States from 6.8

million in 1935 to just more than two million in 2017. Over that same period, there was a comparable increase in the average farm size with farms growing from 155 acres to 444 acres. Market factors and the rising costs of large-scale production were cited as the reasons many smaller family operations got entirely out of farming. Those remaining small farms that failed to maintain maximum-production agriculture would, ultimately, get absorbed into larger operations at deep discounts.

A second primary reason for the disappearance of small family-owned farms was revealed in the 1982 report by the U.S. Commission on Civil Rights. Of our nation's 3.4 million total farmers, only 1.3 percent (i.e., approximately 45,508) are black. Those black farmers currently own a mere 0.52 percent of America's farmland. The report indicated a pattern of discrimination practiced by the federal government and private lending institutions, specifically against black-owned family farms. A pattern of discrimination which prevented them from having equal access to the same credit or crop insurance necessary to sustain or expand their farms as white farmers received.

From 1910 to 2007, black farmers lost 80 percent of their land, which amounted to more than 36 million acres of lost farmlands. This loss of black farmers has overwhelmingly impacted black communities as evidenced through food deserts, severe economic challenges (i.e., poverty rates twice that of rural white communities), and racially targeted environmental pollution (i.e., adverse health effects of fecal waste from hog farms contaminating nearby black and brown communities).

I knew that our family's sugarcane syrup operation alone would not be enough to reverse years of bad agricultural and financial policies leveled against small family farms and black farmers. However, knowing such farms were beleaguered and still disappearing at an alarming rate made the success of our family farm that much more important to me. Uncle Darrell Sr. agreed with my sentiments and advised me to develop a sugarcane farming plan before undertaking any farming activities.

As Uncle Darrell Sr. explained the farm plan serves as a guidepost to the farmer and moves the farming operation along a targeted path. The objective of the farm plan is to help the farm business strategize and take advantage of its strengths, counter its threats, and improve upon its weaknesses. The plan endeavors to identify all resources available to the operation, as well as those resources that are needed but not yet available.

The resources included in a typical farm plan are physical or natural resources (water sources, soil types, land resources), human/personnel resources, animals/crop resources, equipment/facilities resources (tractors, barns, computers, fencing, etc.), and financial resources. As the farming plan developed, I emphasized natural sugarcane syrup as our agricultural product and prioritized our role as stewards of our farm's natural resources. Our goal was for the family farm to become a producer in the North Florida sector of the "Sugarcane Syrup Belt," the area of the deep south renowned for sugarcane syrup production.

In developing the farm plan, my assessment led to a fact-based analysis of our farm's core strengths and weaknesses. Some of our notable core strengths included our natural resources (i.e., ownership of land), equipment/facilities resources (i.e., heirloom cane press and boiling kettle), and human resources (i.e., family cooks with a syrup recipe and experience). However, as mentioned earlier, the syrup's differing taste, texture, and viscosity were not conducive to commercial scaling of the operations. Such individuality appealed to the cottage industry and local farmers' market, but not necessarily the commercial market.

Sadly, the analysis ultimately showed that trying to enter the well-established liquid sweetener shelf space as an under-resourced newbie with little expertise and an inability to commercially scale seemed like a road to failure. No amount of memory lane romanticism could overcome that hard truth about our small, inefficient, cottage industry sugarcane syrup operation. However, inspiration from the upcoming federal farm bill legislation would compel me and my family to try farming a new crop in a new market; industrial hemp.

Part II: Heritage

Heritage (noun): |her-i-tij| Something transmitted by or acquired from a predecessor; something that is handed down from the past, as a tradition.

Sunset overlooking one of the Red Hills fields on Henry Family Farms.

Chapter 4: Obama's 2014 Farm Bill

On February 7, 2014, President Barack Obama signed the 2014 Farm Bill (Agricultural Act of 2014, Pub. L. 113-79,) allowing universities and state agricultural departments to begin cultivating industrial hemp for limited research. Historically, the Farm Bill had been a series of legislation enacted as part of President Roosevelt's New Deal legislation. As a response to the economic and environmental crises of the Great Depression and the Dust Bowl of the 1930s, the original goal of Roosevelt's Farm Bill series of legislation was to stabilize food prices for farmers and consumers, ensure an adequate food supply, and conserve the nation's natural resources.

The Farm Bill now serves as a multiyear package of federal legislation that governs an array of agricultural and food programs. The bill expires every five years but is updated by Congress to comprehensively address agricultural and food issues over the course of the following five years. It has had a tremendous impact in determining what kinds of agricultural commodities are grown, which programs provide crop insurance or subsidies for which foods, and how to ensure healthy food access for many livelihoods. It sets the stage for our food and farm systems.

Obama's 2014 Farm Bill was even more special than its predecessors because it included the legacy of Section 7606, the Legitimacy of Industrial Hemp Research (now referred to as 7 U.S.C.S § 5940). Section 7606 allowed universities and state departments of agriculture to begin cultivating industrial hemp for limited research purposes. Industrial hemp is a botanical class of cannabis and comes from the same species as the psychotropic marijuana plant. According to the version of Section 297A of the Agricultural Marketing Act of 1946 (7 U.S.C. 1621 et seq.) that was current at the time, the term *industrial hemp* means "the plant 'Cannabis sativa L.' and any part of that plant, including the seeds thereof and all derivatives, extracts, cannabinoids, isomers, acids, salts, and salts of isomers, whether growing or not, with a delta-9 tetrahydrocannabinol concentration of not more than 0.3 percent on a dry weight basis."

The 2014 Farm Bill cleared the way for states to license individual farmers to grow the non-psychotropic industrial hemp plant as a regulated agricultural commodity. This legislative approach of using state-sponsored pilot projects to start an industry had been proven successful in Canada in 1998, when the modern Canadian hemp industry was initiated through the government pilot project research. By 2013, the Canadian farmers participating in the pilot projects were making huge profits from industrial hemp. In the U.S., the states of Kentucky, Tennessee, and North Carolina quickly got their pilot programs active after passage of the 2014 Farm Bill, in apparent anticipation of a 2015 growing season.

As a supporter of President Obama, I had emphatically followed his "hope and change" political policies throughout the election. Until now, though, I had not paid much attention to the policies embodied in his Farm Bill. With the information I had amassed while gathering data for sugarcane farming and syrup making, I now considered whether pivoting to a different agribusiness opportunity would be more beneficial. Would an industrial hemp crop as envisioned in Obama's Farm Bill, be successful for a small, family-owned farming operation? If Florida was going to follow the lead of the 2014 Farm Bill and pass laws allowing the cultivation of hemp, then Florida farmers might be on the cusp of a Green Rush opportunity.

I had been a newly licensed city prosecutor during the California Green Rush when their state legislature passed the Compassionate Use Act of 1996 (California Health and Safety Code § 11362.5), legalizing the use and sale of marijuana for medical purposes. Since marijuana was still illegal under federal law at that time, I viewed California's new cannabis laws with skepticism, believing it was not really being used to treat medical disorders. Despite my naive misperceptions about the benefits of cannabis and the Green Rush, I also fully acknowledged that social injustices regularly occurred involving the unequal enforcement of controlled substance laws between different segments of society, specifically in black communities.

Whether it was crack cocaine, opiates, or cannabis, the law enforcement outcomes were unfairly biased against racial minorities, who were more harshly punished.

According to recent American Civil Liberties Union (ACLU) reports, despite similar drug usage rates, black people were 3.73 times more likely than white people to be arrested for marijuana possession. The reports concluded that this disparity was due to racial profiling and bias in law enforcement. Such disparities have not improved but, in fact, have worsened in most states over the last decade.

This reality has had a chilling effect on widespread participation in cannabis-based enterprises throughout the United States by black entrepreneurs. This was definitely a concern for me, my wife, and the other members of my family who were interested in hemp farming. Even though Obama's 2014 Farm Bill had carved out a research exemption for industrial hemp, our small black-owned family farm would be susceptible to additional racial bias and scrutiny if we pursued the opportunity to plant a hemp crop. I prayed and sought spiritual guidance about this business venture. I didn't want to make a rash or hasty decision.

I continued my search for more information about growing hemp. I wanted straight-forward cultivation facts similar to those I had received from FAMU|CAFS regarding growing sugarcane. Regrettably, there was scant information published about farming industrial hemp when the 2014 Farm Bill was approved. California's 1990s Green Rush had produced several cannabis-cultivation books that have since become widely regarded as bestselling classics. However, no matter how informative and helpful those books might be, there is something understated and elusive about properly growing industrial hemp on a commercial scale in an open field.

I began reaching out to farming groups and associations that were already participating in the hemp pilot projects for other states. I requested information about growing the plant, locating seeds, and the necessary farm equipment. The folks I contacted in Kentucky, Tennessee, and North Carolina were helpful but couldn't provide more information than what was already publicly available. They were still conducting their own research and had not finished vetting the vendors in their own states. I again turned to a family member for more information.

My younger cousin, George Jr., graduated high school in Tallahassee in the 2000s, before heading west to California on a football scholarship. He often remarked that the cannabis Green Rush was still underway by the time he arrived in California for college. He said the Green Rush demand allowed him to work for cannabis growers and cultivators during the football off-season.

While watching those growers, he had learned certain cannabis cultivation techniques that would yield maximum canopy coverage as well as a high THC content. He cautioned, though, that an entire crop could be rendered worthless if the grower did not pay careful attention to several factors while cultivating the plant. As an example, he pointed out that the male cannabis plants had to be separated from the female cannabis plants early in the cultivation process. If not, the male plants would pollinate the female plants and the composition of the entire crop would change for the financial worse. In his experience, it was tricky to get the timing right when using these techniques.

Despite these cultivation hurdles, Cousin George Jr. acknowledged that, during his time in California, he had successfully cultivated a few cannabis plants from seed through to harvest. He was certain that his past cannabis cultivation experience would enable him to do the same in Florida with industrial hemp. Hearing about my cousin's previous cultivation experience allayed some, if not all, of my anxieties concerning the understated and elusive cannabis cultivation techniques I sought. Disappointingly, though, my concerns about cultivation techniques would be rendered moot by state politics. It would be another three years after Obama's 2014 Farm Bill took effect before Florida would seriously consider undertaking an industrial hemp pilot project.

Chapter 5: F.S. 1004.4473 and Private Farms

On March 3, 2017, Florida Senators Bill Montford and Bobby Powell filed SB 1726, directing the Florida Department of Agriculture and Consumer Services (FDACS) to authorize and oversee the development of industrial hemp pilot projects at the two state agricultural universities: FAMU|CAFS and the University of Florida Institute of Food and Agricultural Sciences (UF|IFAS). Florida had finally started to establish an industrial hemp pilot project pursuant to Obama's 2014 Farm Bill.

I was delighted to see Senator Montford's name on SB 1726. He sat on the Senate Agriculture Committee and had previously served as its chair. In addition, when he was a Leon County school teacher back in the day, the senator had taught a few of my aunts and uncles at Amos P. Godby High School. I imagined that, with his leadership, SB 1726 would contain a provision that provided an entry point for small family farms to grow industrial hemp in Florida.

The provisions in the final version of SB 1726 stated, "(2)(a) The purpose of the pilot projects is to cultivate, process, test, research, create, and market safe and effective commercial applications for industrial hemp in the agricultural sector in this state." The final provisions further stated, "(4) A university that implements an industrial hemp pilot project shall develop partnerships with qualified project partners to attract experts and investors experienced with agriculture and may develop the pilot project in partnership with public, nonprofit, and private entities in accordance with this section and all applicable state and federal laws."

This was a clear indication that the financial survival of the state hemp pilot project had been intended to depend on private sector-funded research. This dependency on private sector-funding created an opportunity for private farms to participate in the hemp farming research. On June 16, 2017, the final version of SB 1726 was approved by Governor Rick Scott and later became Chapter No. 2017-124 in the Florida Statutes. The table was set for the Florida Green Rush.

Chapter No. 2017-124, codified as F.S. 1004.4473 (Industrial Hemp Pilot Project), set the duration of the hemp research projects at two years, and directed FDACS to oversee and implement the pilot projects researching the growth and cultivation of industrial hemp at FAMU|CAFS and UF|IFAS. Specifically, FDACS was charged with adopting requirements for certifying and registering the sites that would be used by the universities for the industrial hemp pilot projects. FDACS proposed Rule 5B-57.013, F.A.C. (Industrial Hemp Planting Permits) at that time, however, made no differentiation between licensing categories which did little to help the progress of Florida's emerging industrial hemp industry.

In contrast, the 2017 rulemaking conducted by the North Carolina Industrial Hemp Commission for their industrial hemp program, proposed and ultimately approved, farmer-friendly provisions featuring different categories of licenses (for "research only" or "research with intent to market") along with reasonable reporting requirements for THC sampling and compliance, and adoption by reference of Obama's 2014 Farm Bill. In this manner, North Carolina fully embraced Obama's 2014

Farm Bill and demonstrated their commitment to enabling farmers to grow the new crop.

Moreover, F.S. 1004.4473 provided that the eligible universities would work on the industrial hemp research project with a qualified project partner who was a cannabis/hemp subject matter expert, or an investor aligned with interests in the subject matter. The statute further required that the qualified project partner provide "...proof of prior experience in or knowledge of, or demonstrates an interest in and commitment to, the cultivation, processing, manufacturing, or research of industrial hemp, as determined by the department."

Given that hemp had been an illegal substance to cultivate in Florida until the enactment of F.S. 1004.4473, logic directed that there would be no state university that could claim expertise in cultivating the plant without admitting to the commission of a felony. Nevertheless, FDACS proposed rules seemed to repudiate this interpretation and wrongfully viewed private sector farms as adversaries to Florida's industrial hemp pilot project. It appeared that Florida did not intend for its hemp pilot projects to ever reach the croplands of private farms. The essence of the pilot projects envisioned by Obama's 2014 Farm Bill, as documented in the purpose provisions for F.S. 1004.4473, were being circumvented at the state level.

Chapter 6: Hemp Advocates

By October of 2017, I was exasperated at what I felt was Florida's blatant misinterpretation of the intent of the 2014 Farm Bill which included private sector farms in establishing industrial hemp pilot projects. Often during these early days involving legislation and rulemaking for Florida's hemp industry, I found myself as one of only a handful of people in the room advocating for private farmers to have access to hemp. One such hemp advocate who tried to provide positive input to the legislators was a retired mechanical engineer named Bob Clayton.

For a while, Bob Clayton was considered the Florida "Ambassador of Hemp." This was a well-deserved accolade given his advocacy for industrial hemp and his accomplishment of building an entire house made from hempcrete. Bob also published a pamphlet entitled,

"*Industrial Hemp is Five Crops,*" the first page of which boldly states that "...hemp is for farmers...". The pamphlet laid out the necessary fundamentals for a potential hemp industry in Florida. It provided specificity regarding harvesting techniques and retting methods for a harvested hemp fiber crop, along with explanations designed to change the opinions and votes of misinformed legislators who needed to hear the truth about industrial hemp. Additionally, Bob periodically held seminars and workshops where he gathered with other advocates to discuss political strategies to gain greater access to the hemp plant for private growers, cultivators, and manufacturers in order to provide more research and information about the plant's capabilities.

I attended several of Bob's hemp advocacy workshops, with my wife, Chan, or Cousin George Jr. Bob's presentations were always informative and on-point. There was discussion about how the state hemp pilot project statute was being misinterpreted by various state entities. Some attendees discussed how to amend the current hemp laws, while others devised ways to participate in the rulemaking already underway at FDACS. There was a general feeling of disillusionment with the legislative process, and calls for more vocal and demonstrative protest at the state capitol became the rallying cry.

As a result of his workshops, Bob was able to communicate the collective angst of his fellow hemp brethren and sistren in a succinct, productive manner to FDACS, the legislature, and anyone else who needed to hear it. I was pleased to discover other people who also saw the contradictions and hypocrisy in how FDACS was taking virtually no action to implement Florida hemp laws. In fact, it was at one of Bob's hemp advocacy seminars that my wife and I first met one of the executives from the hemp industry company with whom our family farm would eventually partner in the inaugural year of the FAMU | CAFS Industrial Hemp Pilot Project (FAMU | CAFS IHPP).

Chapter 7: The Hemp Farming Plan

As 2017 ended begetting the hopes of new beginnings in 2018, I pressed on and decided to prepare a hemp farm plan similar to the sugarcane farming plan Uncle Darrell Sr. had instructed me to develop. In that earlier farming plan, I reviewed the natural resources, human resources, and equipment/facilities resources available to us for farming sugarcane. To properly assess our then-current farming capabilities for industrial hemp, I would have to adjust the sugarcane farm plan to account for the differences associated with growing a regulated cannabis crop.

Cousin George Jr. had inherited from his father, Great-Uncle George Sr., a 1970s era 135 HP John Deere 4430 tractor. This tractor had a good set of front weights, a solid three-point hitch, a moldboard plow, and a heavy disc harrow. These items comprised the primary pieces of farming equipment available to us for conventional tillage farming. Notwithstanding their age, these pieces of equipment were a strength to our enterprise because they were family owned and in solid working condition.

Cousin George Jr. disc tills a field with the 4430 John Deere tractor.

Additionally, my cousin's access to his father's vacant farmhouse/office would also be important. The farmhouse/office site rests on part of the property where a small livestock watering pond meets the slow rise of a hill overlooked by a large moss-covered oak tree. The building was erected off-grade and featured a partially enclosed, cinderblock carport under the house on the lower side of the hill facing the pond. Unfortunately, Great-Uncle George Sr.

never got to complete construction before his passing. His son, Cousin George Jr., subsequently converted this carport into a basement-style drying facility. The natural resources were next for assessment, which included our soil and water resources.

Our family's farming heritage primarily relied on rainfed watering as a natural resource for watering all of our planted crops. Historically, rainfed farming has been a standard practice for black farmers across much of the South. In fact, farming that depends on rainwater represents about 80 percent of the croplands under cultivation worldwide and currently produces approximately 60 percent of the global food crops.

Rainfed farming had been successful for my ancestors on the loamy and clayey soils found in the Red Hills region because of the soil's ability to store a large quantity of rainfall. Given the backdrop of my family's rainfed farming heritage, I thought the same would be sufficient for watering our five-acre field of industrial hemp. I further imagined that if there was an absence of rainfall, we could water the hemp crop by running a few tripod sprinklers. Given this premise, I didn't attempt to test or evaluate the natural resource of our water any further at that time.

The soil was the other primary natural resource in our farming operation. Uncle Darrell Sr. stressed the importance of knowing our soil composition by having it tested before trying to grow anything in it. He recommended that I test the soil several months before planting to determine the existing nutrient levels and nutrient availability for healthy plant growth. The soil tests would determine pH levels, levels of organic matter, and macro- and micro-elements. I performed miserably at chemistry back in high school and didn't understand many of the terms my uncle was now describing to me. My quest to farm industrial hemp had now turned into a chemistry refresher course that I didn't relish commencing.

The croplands we sampled for soil analysis were those owned by myself, my mother, Aunt Virginia, Cousin Willie Jr., and Cousin George Jr. We each owned adjoining parcels that had been cleared of trees and were ready to farm. The topography of these croplands allowed us to split up their designation into three separate fields: Field Nos. 1, 2, and 3. The total acreage of these croplands was approximately five acres, with 1½ acres in Field No. 1, three acres in Field No. 2, and one-half acre in Field No. 3.

The three fields comprising Henry Farms croplands.

My wife and I took samples from each of the three designated fields and sent them to the UF|IFAS Extension Soil Testing Laboratory. The resulting analysis revealed that, overall, the farm soil had a 5.8 pH with a recommended fertilization treatment of 1.10 pounds of Nitrogen per one-thousand square feet. I didn't understand what these numbers or terms really meant. Uncle Darrell Sr. explained that they indicated the soil was good enough for planting our hemp crop. That's all I needed to know.

I also requested a Custom Soil Resource Report from the Natural Resources Conservation Service (NRCS) of the USDA. This report provided further information about the properties of our soil and advised us on any limitations that might affect our farming uses. According to the report, our Red Hills farmlands have a composition of "Lucy fine sand" and "Norfolk loamy sand" with a clayey substratum. Most importantly, the report's land capability classification for our soil showed suitability for cultivating most kinds of field crops. The land capability section of our report cautioned us to adhere to conservation practices on our farmlands because this composition of soil was susceptible to the hazard of moderate to severe soil erosion. All of these soil reports concluded that our Red Hills soil would be fine for planting hemp.

UNIVERSITY of
UF FLORIDA
IFAS

UF/IFAS Analytical Services Laboratories
Extension Soil Testing Laboratory
Wallace Building 631 PO Box 110740 Gainesville, FL 32611-0740
Email: soilslab@ifas.ufl.edu Web: soilslab.ifas.ufl.edu Phone #:352-392-1950

Landscape And Vegetable Garden Test Report

For further information contact:

Tancig, Mark
Leon County Coop Extn Service
615 Paul Russell Rd
Tallahassee FL, 32301-7060
Tel:
Email:

To:
Jim Jenkins

Client Identification: Parcel A

Set Number: E49776   Lab Number: E126522
Report Date: 22-Jun-18

Crop: Vegetable Garden

SOIL TEST RESULTS AND THEIR INTERPRETATIONS

Target pH: 6.5   *This is the pH at which the above crop will grow at its optimum*
pH (1:2 Sample:Water) 5.8   *This is the pH of your sample in the water medium*
A-E Buffer Value: 7.61   *Buffer pH is the pH of your soil in Adams-Evans Buffer(A-E Buffer). This is done to determine the lime requirement, which will help increase the soil pH to the target pH level desired by the crop.*

Mehlich-3 Extractable		LOW	MED	HIGH
Phosphorus (mg/Kg or ppm P)	117			
Potassium (mg/Kg or ppm K)	71			
Magnesium (mg/Kg or ppm Mg)	52			
Calcium (mg/Kg or ppm Ca)	295			

LIME AND FERTILIZER RECOMMENDATIONS

Crop: Vegetable Garden
Lime: 3.24   lbs per 100 sq. ft.
Nitrogen(N): 0.20   lbs per 100 sq. ft.
Phosphorous(P_2O_5): 0.00   lbs per 100 sq. ft.
Potassium(K_2O): 0.00   lbs per 100 sq. ft.
Magnesium(Mg): 0.00   lbs per 100 sq. ft.

Soil tests results for Field No. 1.

Our human resources were the next item to consider. I didn't know how many people would be needed to grow industrial hemp or how their help would be utilized, but I intended to rely on the other members of my family who were interested in hemp farming. Several of them had already agreed to help plant and cultivate the new crop. However, they were concerned about the unsettled nature of cannabis laws in our nation and the discriminatory law enforcement that came with it.

At that time, in 2018, hemp farming was swathed in a labyrinth of federal laws, state laws, federal agency "statements of principles," and U.S. attorney general memoranda, which increased uncertainty surrounding this agricultural commodity. Regardless of race or ethnicity, this myriad of regulations spawned severe penalties for any suspected noncompliance by farmers with purely innocent intentions. Added to this morass were the documented social injustices against black people caused by misguided law enforcement policies surrounding cannabis. One of my family's most plausible nightmare scenarios while contemplating participation in the FAMU|CAFS IHPP involved a state or local official trying to seize and confiscate our family land on baseless and unproven presuppositions of noncompliance. The impact of that uncertainty and unequal punishment on prospective black hemp farm owners, including ours, was a chill as cold as the Arctic Circle.

To warm-up my hesitant family members, I reminded them that the state-sponsored hemp pilot projects in Kentucky, Tennessee, and North Carolina were launched with legal protections for their prospective farmers. I also reassured them that Florida was certain to provide similar protections for their hemp farmers. However, I had to admit that I could not foresee what those protections would look like.

Nevertheless, I ultimately calmed their fears by pointing out that the 2014 Farm Bill provided the primary protection to farmers by authorizing the pilot projects in the first place. The logic of this arrangement did not escape them, and I enthusiastically completed drafting our simple hemp farming plan. I presented the finished plan to my interested family members, and they received it with warm approval.

Chapter 8: Right Place, Right Time

Later, near the end of the summer, in August of 2018, I attended a hemp pilot project workshop held by UF|IFAS at their Quincy, Florida, extension offices. At this workshop, UF|IFAS explained their vision for their university's upcoming industrial hemp pilot project. They even had a representative from the Kentucky Department of Agriculture make a presentation describing their pilot program, which had been in operation since 2015.

The Kentucky agriculture representative stated that their preliminary production data report for 2017 showed private farmers in their pilot program were paid $7.20 per pound for floral hemp matter. The average per-acre yield for floral matter was 1,024 pounds. That calculated to revenue of more than $7,000.00 per acre for farmers in the Kentucky hemp pilot program. That seemed like a more than decent return on investment to me, though I didn't know any of the input costs. The remaining question was whether UF|IFAS was considering including small private farms in their industrial hemp pilot project in a manner similar to the Kentucky pilot project.

Although participation by small private farmers was not offered at the pilot project workshop, UF|IFAS did provide attendees the opportunity to apply for membership in their pilot project advisory group. This advisory group was to meet two to four times per year throughout the duration of the UF|IFAS pilot project. UF|IFAS stated that they wanted the advisory group membership to represent the diverse interests and stakeholder groups relevant to the hemp industry in Florida. Surely those few slots in the advisory group would be filled with experts and businesspeople from the cannabis, agriculture, and manufacturing industries.

I took measure of my lack of traditional farming experience, and my even greater ignorance of growing industrial hemp. I didn't think mere enthusiasm for the crop alone would be sufficient to garner placement in the advisory group. However, Dr. Oghenekome U. Onokpise, a family friend of Uncle Darrell Sr., was also in attendance at this workshop and suggested that I apply for membership to the advisory group.

Presentation during UF|IFAS industrial hemp pilot project workshop.

Dr. Onokpise, or Dr. Kome as he is known to his colleagues and friends, is a world-renowned agronomist and research scientist who retired after an illustrious teaching career at FAMU|CAFS. Dr. Kome suggested that I apply for membership to the advisory group because I could provide a valuable contribution based on my interest in the alternative crop and my 15 years as a lawyer working with regulatory compliance under federal and state laws. With thanks to Dr. Kome for his encouragement, I submitted my name for consideration to the advisory group. I would rely on Dr. Kome's sage advice time and time again throughout this hemp farming journey.

A few months later, on October 10, 2018, Hurricane Michael made landfall in the Florida Panhandle, with maximum sustained winds of 160 miles per hour. It was the first Category 5 hurricane to strike the Florida Panhandle since Hurricane Andrew in 1992. The extreme winds and storm surge brought about catastrophic damage, flattened homes, felled trees, and caused extensive power outages over a wide swath of the region. My family and I were fortunate that none of us faced any personal injury or major repairs to our homes from hurricane damage.

I checked the fields we had designated for hemp farming. There were no areas of standing water, meaning the rainwater had drained away just fine. Even though there were many fallen trees surrounding our fields, few limbs and little debris had blown onto the cleared croplands. Later that week, when electricity had been restored to our area, I checked my email.

I was pleasantly surprised to discover that I had been selected as a member of the UF|IFAS Industrial Hemp Pilot Project Advisory Group. I was ecstatic to receive the invitation and enthusiastically accepted. My reverie was soon curtailed, however, by the discovery that the UF|IFAS Pilot Project would not involve the use of any private farmers in its inaugural growing season. The growing of hemp in the UF|IFAS Pilot Project would only occur on their university research farm properties.

Despite the clear purpose stated in F.S. 1004.4473, UF|IFAS interpreted the statute in a way that intentionally left out private farmers. I felt as though the rug had been snatched from beneath me once again. Although one of the goals of the advisory group was to provide oversight for pilot project activities, it seemed to be a meaningless exercise. I questioned how effectively an outside group could oversee a project that would only occur on university grounds and only use university resources. It again seemed like the state intended the pilot project to never leave state property or allow hemp research on the croplands of small private farmers. That was until about a week later, when I received an invitation to negotiate (ITN) for participation in the FAMU|CAFS industrial hemp pilot project.

As a review of the ITN provisions would confirm, the FAMU|CAFS IHPP would allow private farmer participation. The ITN defined a "respondent" who applied to participate in the FAMU|CAFS IHPP as a "...company/firm that responds in full to the requests of this ITN and wishes to be considered as a candidate/partner in an industrial hemp pilot project as described herein." This provision made it clear, at least to me, that the FAMU|CAFS IHPP was trying to work with private entities in this research. I informed the other interested members of my family about the university's change in course. We were elated and eager to find out how the university's research project would proceed. I was excited to finally have something in my hands, that might lead to participation in one of the state's industrial hemp pilot projects. However, upon reviewing the details of the ITN, I was shocked to discover that the qualifications FAMU|CAFS required from respondents were quite staunch. I wondered, as I reviewed the requirements once more, if my excitement might have been premature.

Among the listed requirements was that the selected entity must obtain performance bonds, surety bonds, sub-guard insurance, professional liabilities insurance, workers' compensation insurance, commercial general liability insurance, and professional liability insurance. This appeared to be financial security overkill and felt to me like the university was saying, "Small farms need not apply." F.S. 1004.4473 only requires a university to submit proof of such financial assurances if the university research project planned to cultivate industrial hemp in a planting area greater than two contiguous acres. Even that was onerous to me, but not as burdensome as the FAMU|CAFS ITN requirement.

Obtaining the types of financial instruments required by the ITN would be expensive and drain our farm's limited financial resources. Those same resources were needed for the future purchase of seeds, fertilizer, and other inputs to successfully grow our hemp. Despite what I considered unconscionable financial requirements, I continued to prepare our family farm's documents for potential participation in the pilot project. I was still glad that

FAMU|CAFS was providing an opportunity to work with the new crop, even if they weren't making it easy.

Later that same month, I was surprised to receive a call from Ms. Scheril Murray-Powell, whom I previously met at a Florida legislative hemp regulation hearing. Ms. Murray-Powell is an attorney "cannapreneur" who served as the in-house counsel for a privately-owned hemp-derived CBD company. She was leading a partnership application effort to participate in the FAMU|CAFS IHPP and wanted our family farm to join them as one of their farming partners. Ms. Murray-Powell's consulting company, Green Sustainable Strong, LLC, along with another private group serving as the financing partner, had developed an industrial hemp research plan in response to the ITN. The research plan was speculative but practical, and it seemed like good fortune had graced our family farm to be in the right place at the right time.

Cousin George Jr. and I subsequently met with Ms. Murray-Powell to discuss the mutual expectations between her partnership members and Henry Family Farms. The Green Sustainable Strong Group would provide expertise regarding obtaining seeds and cultivating the plants, as well as the financial securities required for the permit. We, as the farming partners, would be expected to provide the equipment and facilities resources, the human labor resources, and the natural Florida resources of our land in which the hemp crop would be planted.

Once we agreed on terms for the research project, I prepared our family's farming partner documents for inclusion in the Green Sustainable Strong Group's ITN response packet. I felt that there was no way the FAMU|CAFS selection committee could leave us out of progressing to the next qualifying round. The principals and titleholders in our group were black-owned entities. Each represented a different sector that would be needed to sustain a fledgling industrial hemp industry (i.e., agricultural, compliance/legal, financial, consulting and marketing). This would be the type of diversity sorely needed in farming generally and the cannabis industry specifically. However, before the FAMU|CAFS IHPP Advisory Board issued its selections, there was a game changer at FDACS for hemp advocates and the pilot projects in the State of Florida.

On November 6, 2018, Democrat Nikki Fried, was elected as the new commissioner to head FDACS in replacement of the term-limited incumbent, Republican Commissioner Adam Putnam. Prior to being elected, Commissioner Fried worked as a government consultant and cannabis lobbyist who advocated for the expansion of patient access to medical marijuana. In the earlier days of Florida's cannabis and hemp legislative debates, she had advocated for some of the same positions as Bob Clayton and Ms. Murray-Powell.

Commissioner Fried took note that the 2018 Farm Bill was being drafted at the federal level by President Donald Trump's administration and would soon have a significant beneficial effect on hemp farming. Therefore, she declared that her FDACS office would move swiftly to approve Rule 5B-57.013 F.A.C. in order to provide specifics for FAMU|CAFS and UF|IFAS to implement the state hemp pilot projects pursuant to F.S. 1004.4473. As a hemp advocate, Commissioner Fried wanted to get a jump on the required FDACS rulemaking process so that the state could adopt its hemp rules within 90 days after the 2018 Farm Bill was expected to pass in December. Commissioner Fried believed that such initiative by her agency would allow private growers to start importing hemp seedlings for planting in time for the 2019 growing season.

With Commissioner Fried now heading the agency, FDACS could fully implement and establish industrial hemp pilot projects through our state agricultural universities and private farms.

On December 4, 2018, the FAMU|CAFS IHPP Advisory Board finally issued its short list of selected respondents. We were heartbroken. Out of the eleven applicants who responded to the ITN, the Green Sustainable Strong Group's response was not one of the seven chosen to advance to the next round. Since the focus of the research project involved farming, I wondered if our part of the response as the farming partner had somehow been deficient. There was no way to tell, though, without any further details.

FAMU|CAFS revised the next phase of their IHPP qualification process, requiring a group presentation with the advisory board evaluation team before admission into the next round. The first-round evaluation had only involved a documentary assessment. This change allowed the university to develop a more detailed inquiry of the respondents before making their selection.

After a quick discussion with the other members of my family who were interested in hemp farming, it was decided that we should look through the short list of the remaining seven groups and try to find another group to join. The proverbial doorway to get into the state pilot project was right here in front of us, and all that we needed was a cannabis industry partner to gain entry. After a cheerless phone call with Ms. Murray-Powell confirming the end of our partnership with her group, I began my review of the entities named on the short-list. A few of the listed entities were small farms that grew specialty crops for niche food interests. Others appeared to be nothing more than company charters with unproven backgrounds growing anything.

However, I immediately recognized one of the companies named on the list. My wife and I had met the executive officer of this particular hemp industry company at a Bob Clayton event some time ago. It seemed like good luck had graced our family farm, putting us in the right place at the right time, once more.

Chapter 9: The Nursery Visitation

While attending a Bob Clayton hemp advocacy event before the ITN from FAMU|CAFS had been announced, my wife and I met a couple of the executives of a Florida-based, hemp industry company. Since the group with whom we had originally partnered in the ITN response did not make it to the shortlist, I reached out to these hemp company executives we had previously met. They remembered us and responded enthusiastically. After some discussion, we agreed to trade site visits. My wife and I would visit their Central Florida nursery site, and their staff would visit the Henry Family Farms site in Leon County.

The hemp industry company is a subsidiary of a large, successful, retail plant nursery business that serves commercial and residential customers in and around the northern part of Central Florida. Painted wooden barrels and other festive planters decorated the grounds of their massive facility. The nursery housed the hemp industry company's business operations, including a 3,000 square foot greenhouse space dedicated to plant propagation and germination. We discussed the various aspects of our mutual interest in the hemp pilot project, as we toured the nursery grounds in a golf cart with the hemp industry company executives.

The hemp industry company's staff acknowledged their cannabis cultivation experience included farm consultations and hemp oil extractions in the states of Colorado, Tennessee, Montana, and Kentucky. They had undertaken a great amount of planning and preparation in anticipation of farming hemp in Florida, whether it was with a state university pilot project or privately after state permits had been issued. As the tour continued, we discussed the mutual expectations for a potential pilot project partnership.

The expectations we had for our industry partner were the provision of financial security for the research project, and expert support in all phases associated with farming industrial hemp, including cultivation, harvesting, and the marketing/sales of the harvested crop. The hemp industry company's expectations for us as a farming partner involved the use of our soil and water natural resources for planting the hemp crop, along with the facilities and human resources necessary to farm the crop. In addition, both FAMU|CAFS and FDACS made it known that state law required them to make regular visits to approved pilot project grow sites to observe, inspect, and test the planted hemp crop. As a result, both state entities would charge pilot project grow sites for the cost of these official visits. The ITN respondents had been informed that these charges would include the employee time and travel expenses to and from the grow site.

The hemp industry company only had one other farming partner, whose croplands were near them in the central part of the state. The distance from FAMU|CAFS and the FDACS plant division to the central Florida facilities of our prospective hemp farming partners would make the travel expenses charged for those required site visits very expensive. Due to our family farm's North Florida location and proximity to FAMU|CAFS and FDACS, our inclusion as a farming partner would lower the travel costs resulting from those pilot project official visits. The hemp industry company agreed that including our family farm would be a win-win scenario for their overall partnership and Henry Family Farms.

The hemp industry company also required their farming partners to obtain industrial hemp seeds and plant clones through them. This was perceived as a strength in the partnership structure. The hemp industry company planned to show their farming partners how to correctly cultivate a THC compliant hemp crop and make their own hemp cultivars for future plantings or sales to other farmers. In addition, they had a hemp processor in their network who would perform oil extractions for their farming partners' harvested crops. That type of access to processing would make marketing and distribution of the harvested crops easier for the farmer.

The nursery visitation concluded on a note of optimism bringing about a tentative agreement to a partnership with the hemp industry company contingent upon the assent of my family members who were interested in farming hemp. My subsequent meeting with those interested family members revealed their elation at having another opportunity to farm hemp in the FAMU|CAFS IHPP. Fully engulfed in the Green Rush fever, I gleefully prepared our portion of the required farming partner documents and submitted them to our new hemp industry partner for presentation in the next round of the FAMU|CAFS IHPP selection process.

Chapter 10: The Farm Visitation

Our industry partner fulfilled their promise to visit us at Henry Family Farms. They reviewed our simple farm plan, examined our facilities, and assessed our capabilities to cultivate hemp. Our attempt to make the best use of available resources without purchasing any new equipment was laudable, in their opinion, but they knew we had no idea what we were doing when it came to cultivating industrial hemp on a farm-scale. The simple farm plan we had developed was based on what we knew, which was growing traditional row crops. However, our industry partner emphasized that industrial hemp was not a traditional row crop and should not be cultivated as such. Instead, we needed a hemp cultivation plan.

One important reason why industrial hemp can't be farmed as a traditional crop is the level of monitoring a hemp farmer must perform on a regulated hemp crop while in the field. Farming a traditional crop doesn't usually require close monitoring. However, the farmer must closely monitor a hemp crop in order to avoid enforcement action resulting from noncompliance with established THC regulations. Such noncompliance is commonly known as growing a "hot crop" and routinely results in the destruction of the entire crop. To prevent this, we needed a cultivation plan developed specifically for farming industrial hemp.

As the industry partner looked over our farm topography and field layout, they estimated we might be able to fit as many as 1,200 hemp plants per acre. That would amount to a total of 6,000 industrial hemp plants if we were to max out the plant count on the five acres we designated for growing hemp. Having reviewed the number of hemp plants grown by other out-of-state pilot programs, I had not imagined we would need more than a few hundred plants total in our inaugural growing season. At least that's what I told my other family members who were interested in farming hemp: "it will only be a few hundred plants." I was staggered by our industry partner's proposition.

The industry partner reminded us that, as one of their farming partners, we would be expected to only use hemp seeds or clones that they acquired for all farming partners. They further explained that requiring this type of exclusivity ensured the right hemp seeds or clones, with the right genetics, were being used from the beginning of the hemp cultivation process. Genetics, in this context, generally means a hemp plant's heritable cannabinoid and terpene traits that can be passed through the plants' seeds or through cloning. Cloning, in this context, is the process of taking cuttings from a mother plant to produce new plants with identical genetics. Our industry partner emphasized that the hemp plant's heritable cannabinoid and terpene traits impacted the plant's potency and character.

Cannabinoids and terpenes are natural genetic chemical compounds found in industrial hemp plants. Terpenes are the aromatic compounds that determine a distinctive flavor and odor for the plant. Cannabinoids appear in the plant as different types, including delta-9-tetrahydrocannabinol (delta 9-THC) and cannabidiol (CBD). Psychoactive delta-9-THC cannabinoid is the primary ingredient in a marijuana plant. The concentration of delta-9 THC in a marijuana plant is the determining factor in whether the plant harbors an illegal Schedule 1 drug.

Delta-9 THC is also present in legal hemp plants, but only in low concentrations of not more than 0.3 percent on a dry weight basis. The primary cannabinoid in hemp plants is non-psychoactive cannabidiol (CBD). Cannabidiol (CBD) is considered the cannabinoid responsible for causing the feeling of anxiety relief hemp has been known to produce. Ultimately, these heritable cannabinoid and terpene traits affect a hemp plant's germination rate, harvest time, yield, and whether its THC levels will remain at or below the legal limit of 0.3%.

After hearing this cannabinoid and terpene explanation from our industry partner, I now better understood the importance of starting a hemp grow with the right genetics. Any previous underestimation I had about the importance of exclusively growing hemp cultivars sourced by our partners was cast aside. In fact, I wondered if this was the elusive cannabis cultivation information of which I had so earnestly sought? Our new partnership in the FAMU|CAFS IHPP would be the vehicle we used to tryout those cultivation practices and test the underlying theories.

Our industry partner continued b[

as one of the farming partners, Henry F;

only be charged half the market price f

According to our partner, the price to tl

$6.00 each when the state began issuing f;

at half price, our cost would be $3.00 e;

calculation of the costs for 1,200 plan

Family Farms as a farming partner woulc

up to $3,600.00 per acre for hemp seedl

me staggered once more.

I was initially indignant about

exclusively use the industry partner's ge:

cost. We hadn't planned on spending tl

seeds or clones. However, I remembe

about how genetics affects the heritab

terpene traits. It would be foolhardy f

such important factors up to happens

when it was possible to inject som(

cultivation requirements and, ultimatel:

The cost of $3,600.00 per acre to

compliance with state law and solid gro

a low price to pay for that certitude. I

indus

to the

I had

that c

ry partner's arrangement for genetics, resigning myself
fact that, if the elusive cannabis cultivation information
sought required us to start with the right genetics, then
st would have to be paid.

Chapter 11: Market Uses

During our industry partner's farm visit, they explained that a good industrial hemp cultivation plan should also include the intended market use for the harvested crop. At the time of that farm visit, there were only three emerging hemp markets for consideration: bast fiber uses derived from the plant's stalk; seed oil consumption uses derived from the plant's seed; and cannabidiol (CBD) oil uses derived from the plant's floral matter or flower. Different cultivation methods are required to properly farm industrial hemp depending on which market use is intended for the harvested crop.

Our industry partner didn't want us to grow hemp for the bast fiber market until fiber processing operations were more readily available in Florida. They wanted our farm to consider growing for hemp seed exclusively. Not for the market use of seed oil consumption, however. Instead, they wanted us to grow for seed stock to use in our own future plantings, as well as for sale to other hemp farmers.

They offered two reasons for this suggestion. The first was SB 1020, a senate bill recently filed in the Florida legislature. Among other things, this bill would create the upcoming state-wide hemp program and require licensees to only use hemp seeds and cultivars "...certified by a certifying agency or a university conducting an industrial hemp pilot project pursuant to [F.S.] 1004.4473." In our industry partner's opinion, this law would allow us, as an approved grow site, to produce certified seed or cultivars for licensed hemp farmers throughout the state.

By their estimate, our five acres could produce approximately one million seeds. There wasn't any solid data or information about the future demand for hemp seeds or clones, but one million seeds sounded massive. If we sold all one million at just one cent per seed, that would amount to $10,000.00. That type of revenue would help fund the needed upgrades and expansion we anticipated for our small farm's future grow seasons. This was definitely an option to consider.

The second reason to grow for seed stock was the topography of Field No. 3. Surrounded on three sides by a thick tree line, this field was physically separated from the other planting fields practically making it an isolated nursery.

That meant that, even if we didn't want to grow exclusively for seed development on all our fields, we could grow for seeds only in Field No. 3 while still growing for flower production in the other two.

The biggest potential problem, they surmised, we might encounter was pollination. If our feminized hemp plants in Field Nos. 1 and 2 became pollinated from any nearby male hemp plants, the female plants would then only develop seeds instead of flowering buds. Finding any male plants in Field Nos. 1 and 2 would be a daunting task to begin with, as there would literally be thousands of plants in our fields. Moreover, any seed development caused by male plants would have to be carefully removed by hand so as to not disturb overall flower production of the crop.

Regardless of the difficulty involved in completing the task, we were willing to undertake the arduous, plant-by-plant assessment in order to ensure the viability of our flower production. This was similar to what Cousin George Jr. had forewarned in our earlier discussions about his experience working with growers in California. I now wondered if this was the elusive cannabis cultivation information which I had sought, and not the aforementioned practice of using the right genetics from the beginning of the grow?

I revisited the notion that our new partnership in the FAMU | CAFS IHPP would be the vehicle through which we would tryout these cultivation practices and test the underlying theories. How we dealt with events during the flowering phase of cultivation would eventually determine the content and quality of our harvest. After a talk with the other interested family members, we agreed to follow our industry partner's cultivation plan. We would grow for seed production in Field No. 3 and flower production from which to derive CBD oil in Field Nos. 1 and 2.

Chapter 12: The Hemp Cultivation Plan & Timetable

The conversation with our industry partner moved on to the new timetable established for the hemp cultivation plan. Depending on genetics, the timetable for an industrial hemp crop from seed to harvest can be anywhere from 180 days to 224 days. If germination of the hemp seeds is handled through greenhouse care, growing those seedlings to maturity might shorten the harvest timetable to as little as 90 days for certain varieties of industrial hemp.

Our industry partner emphasized that the time of year during which industrial hemp is planted outdoors can make a significant difference in the crop's ultimate yield. Once the time of year for planting has been determined, the cultivation plan timetable can be created. The timetable lists the primary events that will occur during the grow season, including the planting day, the plants' transition into their flowering event, and the harvest date. These are the times when the farmer must be most vigilant handling their regulated hemp crop.

Our industry partner's cultivation timetable set planting day to occur in mid-March, due to the photoperiodic nature of industrial hemp. This photoperiodic nature means that the cannabis plant has a physiological reaction and developmental response to the relative lengths of light and dark periods to which it is exposed. As a result, cannabis plants, like industrial hemp, have two growth phases, a vegetative phase and a flowering phase, each requiring different periods of darkness and daylight.

Hemp farmers want their plants to remain in the vegetative phase as long as possible in order to promote plant growth and yield a larger crop. Industrial hemp plants in a vegetative phase require about 16 hours of sunlight to remain in that phase. A mid-March planting date ensured our crop would still be in the vegetative phase in the field during the summer solstice. This exposure to the summer solstice sunlight in June, when the sun travels the longest path through the sky, providing the most daylight for any day that year, would help our plants produce maximum yields when later harvested.

When the fall equinox of late September approached and the daily sunlight levels decreased to about 12 hours or less, our plants would transition into the flowering phase. In the flowering phase, the reproductive organs of a hemp plant begin developing. The buds from which the cannabidiol (CBD) oil is extracted bloom.

Our discussion pressed on to the final timetable item: harvesting. This was the one area of our industry partner's cultivation plan where it honestly felt like the entire discussion was premised on, "if you get this far." It was a discouraging presumption, given the steeplechase of tasks we had to accomplish to be successful. I tried to stay optimistic, but the deluge of information I had received from our industry partner during this visit had turned our pleasant family farming journey into the "Twelve Labors of Hercules."

As part of the harvesting component from our industry partner's cultivation plan, we decided to examine Cousin George Jr.'s vacant farmhouse/office as a future storage and drying location for the harvested crop. Everyone present for our industry partner's farm visit agreed that, with a few modifications, it would meet our needs. We decided to perform a more thorough assessment of the facility closer to harvest time. Having completed that discussion, we now focused on the cultivation plan's reliance on drip irrigation.

Our industry partner preferred using a drip irrigation system to water the industrial hemp crop. Drip irrigation is a crop watering method that allows precisely controlled applications of water to drip slowly near a plant's roots. The system is comprised of a network of valves, pipes, tubing, and emitters. Healthy cannabis plants can grow root systems up to three feet in length, making a drip irrigation system focused on those roots more efficient and productive than a tripod sprinkler system.

A drip line irrigation system would also make it possible to fertilize the crop by directly injecting liquid fertilizer into the irrigation system. This is called fertigation, which our industry partner also preferred. A chemical injection device connected to the irrigation system would deliver a liquid fertilizer through the system. They suggested we either build a do-it-yourself chemical injector using PVC tubing or purchase a commercial-scale chemical injector from their network of vendors. If installing an irrigation system didn't present enough of a learning curve, then adding a do-it-yourself fertigation system certainly put it over the top.

I thought back to the early expectations I'd had for watering the plants on our hemp farm. I had idealistically imagined using the agricultural methods of my forefathers to grow this modern-day, regulated crop. In hindsight, it was an uninformed assumption. A modern-day, regulated crop needs modern-day cultivation techniques to produce solid yields. I suspected that my forefathers' rainfed watering tradition might be too inconsistent and unreliable for the requirements of industrial hemp.

We were fortunate to have two working water wells available for irrigation in Field Nos. 1, 2, and 3. The challenge we had to overcome, though, was that the residential water well located between Field No. 1 and Field No. 2 was too far away from either field to properly connect an irrigation system. At some point before installing the drip irrigation system, more than 150 feet of PVC pipe would first have to be trenched from the well to each of those fields.

Neither I, nor any of the other interested family members had ever installed or worked with drip irrigation systems before. Despite this being another Herculean trial, I wouldn't allow it to become the reason for our failure.

Our industry partner concluded their farm visit with the expressed opinion that we had the resources and capabilities to successfully farm industrial hemp. Their well wishes and optimism gave us high hopes for future success and quelled my rising anxiety.

Part III: A New Tradition?

Tradition (noun): |trə-ˈdi-shən | An inherited, established, or customary pattern of thought, action, or behavior; a social custom.

Henry Family Farms hemp planting permit.

Chapter 13: The Starting Line

During a March 2019 FAMU|CAFS Board of Trustees meeting, our industry partner was chosen as one of the "qualified respondents" to participate in the FAMU|CAFS Industrial Hemp Pilot Project. Our partnership group was ecstatic to have finally reached what we considered to be the starting line for farming hemp. For a moment, we even thought we would be able to start growing hemp according to our 2019 mid-March timetable. However, FDACS still had to complete their planting permit process before we could obtain any hemp seeds or genetics. The permitting process for the hemp pilot project was new to FDACS, slowing their administrative processing times. Before we knew it, May 2019 was upon us. We were weeks away from the summer solstice that would occur on June 21st and were not even close to having plants in the ground.

As we awaited issuance of our permit, the FAMU|CAFS research farm manager came out to visit our farm along with a FAMU|CAFS staff member and a Leon County extension agent. Our industry partner also attended this visit, making the meeting feel more like a pilot project collaborative effort on our farm than any other time before. During this meeting, the collected experts determined that Field No. 1 should have 25 rows running east to west, Field No. 2 should have 17 rows running east to west, and Field No. 3 should have 25 rows running north to south. FAMU|CAFS further determined that they wanted to designate Field No. 1 as the primary demonstration field. We were all in total agreement with these determinations.

The FAMU|CAFS contingent then assessed our 30-year-old, residential water well pump, which we planned to use to water Field Nos. 1 and 2. The 175- to 200-foot-deep well was powered by a 1½ horsepower pump that produced an estimated water flow of five gallons per minute. The assembled contingent of experts concluded that despite the pump's age, it might survive the grow season if a drip irrigation system was installed which required less water demand. Installation of a T-valve lever at the well was also recommended to allow the irrigation system to be separately cut on and off. These measures seemed reasonable and dovetailed with the recommendations from our industry partners.

The contingent of experts ultimately sketched out the preliminary design of an irrigation system for use in Field Nos. 1 and 2. This system would produce an estimated water delivery rate of around 30 gallons a minute per acre. That would be a tremendous upgrade over the inconsistent, rainfed watering system we had planned.

Field No. 3, on the other hand, did not require as much planning because its designated water well was solely dedicated to serving the water uses of that field. Since this well had a depth and horsepower similar to the well supplying Field Nos. 1 and 2, it's irrigation system also had a similar water delivery rate of more than 30 gallons a minute per acre. This was fine with us.

The visit to our farm concluded with optimism from the collected agricultural and cannabis experts. In their opinions, Henry Family Farms was only a few upgrades away from having the resources and capabilities to successfully farm industrial hemp.

Chapter 14: Genetics

In July 2019, FAMU | CAFS and our industry partner were issued an industrial hemp planting permit to cultivate industrial hemp on three acres of croplands at Henry Family Farms in Leon County, FL. FAMU | CAFS and our industry partner were named on the permit as the state authorized permittees. Henry Family Farms was not listed as a permit holder because the financial security for enforcement accountability fell on the shoulders of our industry partner as the named permitee. Moreover, we were only approved to grow on a smaller portion of the acreage than had originally been requested. Despite our seemingly invisible permit status and diminutive hemp farming acreage, my family and I were proud to be designated as a state-approved industrial hemp pilot project grow site and for the opportunity to participate in the pilot project.

The late issuance of the FDACS planting permit led us at Henry Family Farms to believe that the first chance to plant hemp would not occur until the Spring 2020 growing season. We believed that to be true, because the summer solstice had passed, and it was now too late for us to plant and properly cultivate any hemp according to our cultivation plan. Our industry partner, though, had already considered a way to salvage the 2019 grow season.

At that time, Kentucky, at three state lines and up to ten hours one-way, was the closest location from which a Florida hemp grower could legally obtain hemp genetics. Our industry partners already had a Kentucky vendor prepared with a few varieties of seedlings ready for transport to Florida.

We chose the Cherry Citrus variety from the vendor's stock based on its genetics. The cannabinoid profile for the variety described it as producing hearty plants containing high levels of cannabidiol (CBD). That high level of CBD is what we needed when growing hemp flower for oil extraction. Our industry partner set a new planting day for the end of September, with harvest to occur by the end of the first week of December. For such a shortened 2019 grow season to be successful for us, it would be necessary to germinate the seedlings and give them several weeks of growth before transplanting them into our fields. The Cherry Citrus seedlings would then have nine to ten weeks in our soil. This would be enough time for them to thicken, grow, and produce flowers.

Next, our industry partner turned their attention to developing a transport plan to get the plants safely back to Florida. Hemp plants are meant to be immovable in soil until harvested. Transporting them up and down highways is not a preferrable cultivation practice. In fact, the plants would undergo a considerable amount of stress during such a journey.

Since the part of the trip involving the plants would take approximately ten hours, the photoperiodic nature of the tender hemp seedlings made it necessary to transport them in an enclosed, refrigerated tractor trailer with multiple levels of lighted shelving. The refrigeration would keep the seedlings from cooking in their trays before reaching a Florida field. Lighting was necessary on each shelf to keep the seedlings in a vegetative growing phase. Without adequate lighting during the long trip, the young plants might prematurely flip over into flowering phase, which would be disastrous.

In addition to worrying about the plants surviving their ten-hour transport, we also had to consider the uncertainty of our nation's cannabis laws. For example, in May 2014, the DEA seized 250 pounds of hemp seeds en route to the University of Kentucky's Industrial Hemp Pilot Project from Italy. This unwarranted action carried ominous reverberations throughout the industrial hemp industry. None of us in the FAMU|CAFS IHPP wanted to pick a legal interpretation battle with the DEA regarding the transport of industrial hemp. We had already set our new planting date and undertaken many preparations. We wanted those seedlings here in Florida without any law enforcement interdiction.

With FAMU's approval, our industry partner secured the required transportation documents, and made the trip to Kentucky. During the ten-hours back, I anxiously watched the news, dreading each weigh station stop they would encounter with the seedlings. The steeplechase of hurdles that confronted our grow seemed never-ending. Despite our partner's use of a refrigerated trailer, I imagined the seedlings stored in that cavernous metal vessel would bake like cookies in an oven, as the average daily temperature on the roads during the trip climbed to highs of 95°F. We let out a collective sigh of relief when our hemp seedlings arrived at the Central Florida nursery.

On August 1, 2019, our industry partner transplanted more than 10,000 hemp seedlings into the Central Florida fields of their other farming partner. This planting immediately made the FAMU | CAFS Industrial Hemp Pilot Project (FAMU | CAFS IHPP) the largest cannabis research project in Florida at that time. Henry Family Farms was proud to be a part of such an historic undertaking, and we looked forward to our own September planting date with enthusiasm.

My wife and I visited the seedlings shortly after their arrival at our industry partner's nursery facility. Our partner provided greenhouse care to the more than 1,500 seedlings bound for Henry Family Farms. Greenhouse care would make them sturdier and better able to endure the rigors of the Florida sun and involved daily watering and cultivation in one-gallon pots for approximately eight weeks before their transplant. Approximately three weeks after arriving at the greenhouse facilities, though, the seedlings prematurely transitioned into a flowering phase. This was a big problem.

My visit with the hemp seedlings upon their arrival at the Florida nursery.

When hemp plants switch to flowering phase, they stop getting bigger and stronger. Our seedlings had to be switched out of the flowering phase and back into the vegetative phase for them to gain size before being transplanted into our fields. By means of nothing short of horticultural wizardry, our industry partner managed to get the seedlings to switch back into vegetative phase. Afterwards, they shone 6,200-watt portable construction site lights on the seedlings each evening from dusk to dawn to keep them in that vegetative phase. The seedlings continued to grow, but the premature transition to the flower phase ultimately stunted their overall size.

Members of the Henry Family Farms cultivation team at the plant nursery tending to our hemp seedlings.

Henry Family Farms would dispatch a cultivation team to our partner's nursery facility several more times while the seedlings were in greenhouse care. During those visits, our cultivation team received training from our industry partner's staff on how to properly prune and cultivate the hemp seedlings. With that many seedlings to account for, it certainly appeared that we would satisfy the farming maxim of growing enough of a crop in order to have enough of it to make sound decisions about using the crop after harvest.

Chapter 15: Preparations

After making that first nursery visit, I threw myself into completing our farming preparations. Field No. 1 was approximately 185 feet west of the water well, while Field No. 2 was approximately 270 feet south of the well. In order to irrigate those fields, I needed to trench those distances on each side of the well and connect two-inch-wide PVC pipes in those trenches to carry the water.

The water well supplying Field No. 1 before trenching.

I rented trenching equipment and dug a 24-inch-deep, four-inch-wide trench to each field from the well. My wife helped me assemble the two-inch PVC pipe from the well to the point in the field where it would connect to the irrigation system. The water pump upgrade and trenching of the PVC piping out to Field No. 1 and Field No. 2 made it possible for us to achieve the estimated water delivery rate of 30 gallons a minute per acre. It took a few days to accomplish, but I was satisfied with the final result. The only plumbing-type work I had performed up to that point was changing the U-joint under my kitchen sink. Now, my wife and I had successfully connected the future hemp fields to a consistent source of fresh, well water.

The water well supplying Field No. 2 after trenching.

The preparation of our soil beds using conventional tilling was the next task to complete. Cousin George Jr. would accomplish this task using his tractor. Primary tillage passes using his moldboard plow would loosen the compacted, clayey consistency of the soil and allow the penetration of air, water, and nutrients. This would be followed by a few secondary tillage passes using the disc harrows to create a soft, suitable seedbed. This process ensured that, prior to our planting day, the soil was well turned over and had a fluffy texture ready to receive the tender hemp seedlings.

The centers of our row troughs were five to six feet from each other. We wanted this much space between them to allow easy passage, promote air circulation, and reduce the incidence of disease among the plants. In order to accurately gauge the amount of drip irrigation equipment needed for installation on our three-acre hemp grow, we had to calculate our field dimensions, determine the number of rows we would plant, and the number of plants in each row. The calculations worked out to 25 rows up to 270 feet long with approximately 25 plants each in Field No. 1, 17 rows up to 500 feet long with approximately 40 plants each in

Field No. 2, and 25 rows up to 250 feet long with approximately 30 plants each in Field No. 3.

We developed a plan to hand-dig the 1,220 plant holes. I thought back to one summer when I was a teenager and Granddaddy Sam instructed me to erect 50 post holes, each 24 inches deep, for a new hog pen fence. Even when I was much younger in age and stronger in muscle, the job of digging 50 post holes was exhausting. Here I was again, like some mad man driven by a fever, preparing to dig 1,220 holes by hand for the sake of our hemp seedlings. It was another Herculean labor to complete.

Fortunately, several family members and friends volunteered to join our cultivation team and help dig the necessary holes for the seedlings. Besides myself, notable members of the assembled cultivation team included my wife, our sons, Isiah and Jordan, my mother and her friend Ms. Portia, Aunt Virginia and her granddaughter, Cousin Justice, Cousin George Jr. along with his mother and some of their friends, Cousin Jim Jr., and my wife's parents, Mr. Gene and Ms. Nancy.

In fact, several members of the cultivation team had previous experience handling large numbers of plants. My father-in-law, Mr. Gene, retired from the landscaping division of the state university system after more than 30 years of working with plants and horticultural activities. The family of my mother's friend, Ms. Portia, had previously owned a landscaping and nursery business in Thomasville, Georgia. She grew up working with and around many species of plants. Such experienced backgrounds in our cultivation team members were invaluable. I was elated that the strength of our human resources was coming through as outlined in our farm business plan.

The cultivation team members measured the distances between the plant holes and marked the locations for excavation. Team members then split into two-person crews that worked on a single row. With one set of post-hole diggers per crew, each team member took a turn digging while the other rested. Each crew developed a rhythm to their labors as they became more accustomed to the process. The cadence of the two-person crews working together brought back memories of my uncles, working in unison, carrying melons from the field to Granddaddy's truck. The work of industrial hemp farming invoked nostalgic feelings, harkening back to a part of my family's farming heritage.

After excavating the plant holes, we were ready to install the irrigation system. A medley of equipment and parts would be used, including blue PVC lay-flat hose, tubes, emitters, connectors, O-rings, and drip tape. Our installation required connecting the blue PVC lay-flat hose to the two-inch pipe in the uncovered trench dug from the well to Field Nos. 1 and 2. When that connection was completed, the drip tape was then connected to the blue PVC lay-flat hose, row-by-row, with an array of tubes, connectors, and O-rings.

Irrigation system drip tape and emitters rated at 0.5 gallons per hour were chosen because they would release water to the ground slowly allowing for the best penetration into the red clay soil.

The installation was slow and methodical as we gained familiarity with the equipment. It seemed that incorrectly connecting one piece ensured another would burst free and erupt like the Old Faithful geyser. Quitting was not an option. We had come too far to let learning a new skill become our stumbling block. Our crop wouldn't stand a chance if this irrigation system did not work.

An overwhelmingly complex problem arose as we installed the irrigation system in Field No. 2. There seemed to be no end to the number of gushing water leaks that erupted from that set of irrigation equipment. This was upsetting to us, as Field No. 2 was our largest parcel, and we had expected to fill it with the most plants. Ultimately, the problem persisted and prevented us from completing the installation for almost half of that field. We were devastated.

Without irrigation for that portion of Field No. 2, we would be forced to rely on the inconsistency of rainfed watering or the haphazard spray from tripod sprinklers. Neither rainfed watering nor tripod sprinklers would support the $3.00 per seedling price we would have to pay for the 500 plants destined for transplant in the non-irrigated part of Field No. 2. The $1500.00 cost for the seedlings would be a gamble on whether rainfed watering or tripod sprinklers would prove successful. Begrudgingly, we decided to only plant the part of Field No. 2 equipped with the working irrigation system. This time, the trial had defeated Hercules.

We completed installation of the irrigation system and successfully tested each of the zones on the remaining fields. We were elated to find no more leaks. It did not escape us that this irrigation system would serve as the lifeline for the seedlings when the brutal Florida sun beamed down on them in the coming weeks. The irrigation system had to work properly, or more than 1,000 seedlings were likely to perish. Overall, I believe this upgrade to our water resource system was the most important preparation we made for farming industrial hemp.

With the irrigation system installed and functional, we needed a transport plan to bring the seedlings to our fields in the Red Hills of Leon County. Our industry partner had previously established a transport plan for movement of the seedlings from Kentucky. Those seedlings suffered a considerable amount of stress during that travel. Another trip could further damage them if done incorrectly. A refrigerated trailer was not necessary this time, due to the shorter transport distance between Central Florida and Leon County. However, multiple levels of lighted shelving would still be needed in the trailer to address the photoperiodic nature of the hemp seedlings.

The plan was to load the seedlings into the trailer at the nursery in the cool early morning hours before the break of dawn. The trailer would arrive at our farm a few hours later, where the assembled cultivation team would unload the seedlings into a staging area, that would be intermittently watered by a tripod sprinkler. FAMU|CAFS had been notified of the expected planting day so that interested members of their staff could be in attendance. I personally invited Dr. Kome to attend.

The evening before our plants arrived, I backed my truck up to Field No. 1 and sat on the tailgate overlooking the rich red soil that was now pock marked with more than one thousand plant holes. I did not start this journey with the intention of farming hemp, and there would be no turning back when those seedlings hit the ground. The outcome of the grow would rely on deft farming skills, which I was sorely lacking. Our family farm would either grow a marketable cannabis crop or a bunch of felonies.

There I sat, having spent thousands of dollars on hemp genetics, awaiting delivery of more than a thousand hemp seedlings I intended to cultivate through to harvest. I was physically drained, yet giddy with Green Rush eagerness at the thought of finally undertaking this historical activity on our family farm. Even though we had successfully traversed many stumbling blocks and obstacles, I quietly resolved to myself that our next hemp grow would be different.

Chapter 16: Planting Day

I awoke early on the morning of September 27, 2019, excited with Green Rush fever about our hemp planting day. The weather was partly sunny with an overall high of 95°F, and average humidity of 53 percent. The lunar projection had the moon waning from a balsamic phase into a new moon phase that would begin the next evening on September 28. Given the weather forecast, we chose to erect a few comfort tents and gather several water coolers filled with ice and water bottles to prevent any heat related health issues among our cultivation team members.

The cultivation team members who had previously excavated plant holes, thankfully, volunteered to return for this important event, along with other invited close family friends. A contingent from FAMU|CAFS attended to represent the university. This assembly of human resources coming together for the planting day event harkened back to the happy communal gatherings for our family's "cane grindings."

Folks arrived in high spirits, excited to see more than a thousand cannabis plants being transplanted into our Red Hills soil. This would be the first time in history that a black-owned family farm would legally plant an industrial hemp crop in Florida. All assembled parties could feel the momentum of Green Rush energy permeating the air. It would focus our actions and drive our labors throughout that humid North Florida day.

The shelving system inside the enclosed trailer built to ensure the safe transport of the delicate seedlings.

107

Our industry partner arrived later that morning driving a seven-foot-long, dual-wheel, enclosed trailer containing 1,220 Cherry Citrus industrial hemp seedlings. When the trailer door opened, we were overwhelmed with the pungent smell of cannabis mixed with the sweet scent of cherries. The delicate seedlings arrived in good shape with very few casualties resulting from their transport. An assembly line process was quickly established to remove the seedlings from the trailer, stage them in an open area near the fields, and intermittently water them with a tripod sprinkler while they awaited transplant.

Seedlings were removed from the trailer and staged in an open area near the fields. While awaiting transplant, the seedlings were watered by a tripod sprinkler.

A set of best practices were developed and adopted by cultivation team members to properly transplant the seedlings. These best practices included thoroughly watering the soil of the plant hole before transplanting the seedlings, carefully removing the seedlings from their pots so as not to separate and disturb the root structures, digging the plant hole deep enough to set the entire seedling root structure underground without any cramping, and pressing the soil firmly onto the top of the roots and covering them with more soil to prevent unwanted air pockets in the ground.

Cultivation team members Mr. Gene, Ms. Nancy, and their daughter, my wife Chan, discuss best practices for transplanting the hemp seedlings.

Seedlings were transplanted by hand at four inches to six inches in depth, where the root masses could stay moist but not become overly saturated. The stems of the plants appeared strong, angular, and branched, with an average plant height of nine to 12 inches. The leaves of the plants measured an average of two to three inches, with new leaf formations appearing on the stems and lateral branches.

The seedlings for Field No. 1 were transplanted the same day they arrived on the farm. Due to the aforementioned difficulties with the irrigation system in Field No. 2, the seedlings for that field remained in their one-gallon pots, and were watered by the tripod sprinkler until their transplant was completed the next day. Ultimately, 796 seedlings were transplanted into Field No. 1, and 424 seedlings were transplanted into Field No. 2.

*Cousin George Jr. plants the first industrial
hemp seedling in our family soil.*

A few weeks after getting our plants for Field Nos. 1 and 2, our industry partner sent us an additional array of 822 hemp seedlings. These seedlings comprised several different varieties for transplant into Field No. 3. We had agreed to designate Field No. 3 to grow hemp for seed development, and our partner donated the plants to us for this purpose. These seedlings also arrived in good shape with no casualties. We staged them in an open area near the fields, in a fashion similar to how the other seedlings had been handled. The cultivation team quickly transplanted these seedlings into Field No. 3.

Warning signs were soon erected designating our three fields as part of the FAMU|CAFS Industrial Hemp Pilot Project. We finally felt that we were officially industrial hemp farmers in the state pilot project. After all seedlings had been transplanted, I walked the field rows to have a closer look. The sun was setting in the early evening hours of the seedlings' first night in our family's soil. I was fascinated to see so many hemp plants in the ground. They were visibly indistinguishable from marijuana plants. I was momentarily overcome with the scope of this venture, hoping that I had not made a huge mistake.

The warning sign for Field No. 2.

There is no better place to witness breathtaking sunset views than from the edge of our Red Hills croplands, which were now full of aromatic green hemp. That evening view seemed to wash away my fatigue and refresh my spirit for whatever Herculean challenges I would confront the next day. Despite having plants in the ground, our family farm had not yet reached its goal of finding a profitable agricultural business opportunity that utilized resources existing on our family farmlands. We still needed to determine if this venture could achieve that end. The primary unknown variable in our equation was cultivating these highly regulated cannabis plants.

Chapter 17: Cultivation Data

Article I: Methodology

In order to keep track of our plants and accurately record collected plant data during the research project, we numbered the planted rows in each field with row marker tags. Using small, orange ground marking flags, we also numbered 20 randomly selected plants in each field for further individual observation. This numbering system allowed us to numerically identify selected plants on a particular row in a specific field.

For the shortened nine-week growing season, we collected and recorded atmospheric statistics that covered the prevailing weather conditions. We also collected and recorded data regarding any incidents of weed pests or plant diseases that arose. Weed pests are described as any plant that is growing out of place or where it isn't wanted. Plant diseases include fungi, bacteria, viruses, and other unwanted microorganisms. Early identification was important for mitigation and removal of either malady, since their competition with our hemp plants for water, nutrients, sunlight, and root structure space would ultimately reduce the hemp crop yield.

Data was also collected directly from the plants in Field Nos. 1 and 2. The seedlings' performance in the areas of new growth, flower formation, and overall plant viability were all recorded. In addition, we monitored the concentration of delta-9 THC in our hemp plants by collecting plant data on a weekly basis commencing from the plants' transplant date. This weekly data collection tracked the amount of time our plants had been in the ground and, thereby, exposed to sunlight.

As mentioned earlier in these pages, the photoperiodic nature of industrial hemp means that sunlight makes the plant produce more of the regulated cannabinoid delta-9 THC. Failure to properly monitor the hemp plant's time in the soil makes the crop susceptible to becoming a non-compliant "hot crop," containing a high concentration of delta-9 THC above the allowed 0.3% legal threshold. A weekly lab testing schedule was designed to mitigate this risk, ensure THC compliance, and gain data on the Cherry Citrus variety's performance in our Florida soil.

To carry out the weekly lab testing schedule, we clipped and dried a sample of flower measuring four to six inches in length from a randomly selected hemp plant in the field. These samples were submitted to a private third-party lab for THC compliance and potency analysis proceeding from the fourth week in the ground through until harvest. These THC compliance and potency lab analysis results are also included in these pages.

In each of the following articles in this chapter, the number of weeks from initial transplant that the plants have been in the ground exposed to sunlight will be reflected in the article heading that precedes its corresponding atmospheric and plant condition data table. For instance, the heading "Article II: Week One Results" describes the cultivation results collected on the first week after transplant of the seedlings. These article titles are consecutively numbered from the first week after transplant through to harvest. The atmospheric and plant condition table for each corresponding week after transplant is entitled "Data

Recorded [*followed by the date*]," denoting the day on which the data was collected.

Lunar phases were also noted, as it is a well settled fact among farmers that phases of the moon cycle have certain beneficial effects on any plant growth. Farmers have used the different phases of the moon to guide the planting and harvesting practices of traditional crops for ages, believing that the same gravitational pull of the moon that causes the rise and fall of tides can also affect the amount of moisture in the soil.

I remember Granddaddy Sam's weathered copy of the Old Farmer's Almanac lying on a table in his farmhouse. My librarian mother and school administrator father raised me to love reading, but I didn't see much use in that old farming book. It contained a bunch of tables, charts, and ads for farm equipment, which were of little interest to me as a child. Throughout the 2019 industrial hemp grow season, however, I found myself repeatedly relying on the Old Farmer's Almanac meteorology forecasts and lunar projections. The lunar projection data provided in the pages of this book was collected with the hope it will assist in identifying a correlation between the phases of the moon and the primary events involved in farming industrial hemp.

Finally, the data collected from the several different varieties of hemp seedlings transplanted to Field No. 3 concerned our industry partner's proprietary seed development information. That proprietary information was not meant for publication and will not be further discussed in these pages.

Measuring a hemp plant in the field.

Article II: Week One Results

Data Recorded Oct. 7, 2019		
Avg. Temperature	High 97°F	Low 70°F
Avg. Humidity	69%	
Avg. Plant Height	10" to 14"	
Avg. Plant Width	3" to 4"	
Plant Survival Percentage	99% survivability *20 plants died for undetermined reasons.	
Lunar Projection	Waxing from a crescent phase at 48 percent illumination, going into a first-quarter phase on October 5.	
Weed Pests	Nut grass (i.e., nut sedge, Cyperus rotundus) began growing in small patches between the rows and around the plants in both fields.	

Our plants went into the flowering phase one week after transplant; approximately nine weeks (63 days) after their greenhouse care had commenced. Shortly after a hemp plant switches to the flowering phase, it undergoes a burst of rapid growth known as the "stretch." This stretch is a growth spurt that gives the plant enough size and strength to support the coming flower buds. The stems of our plants were strong, angular, and branched, with robust signs of growth.

We were still working through our newly established processes during that first week. A farm manual was developed containing our standard operating procedures, copies of our permit, operational checklists for best cultivation practices, chains of custody, and certificates of analysis (COA) for each hemp variety on site. The COA is a report from an accredited laboratory providing a profile showing quantities of various cannabinoids, terpenes, and contaminants in the analyzed hemp plant's genetics. It is a document every hemp farmer should be familiar with and have at every transaction involving a cannabis plant or cannabis plant by-product.

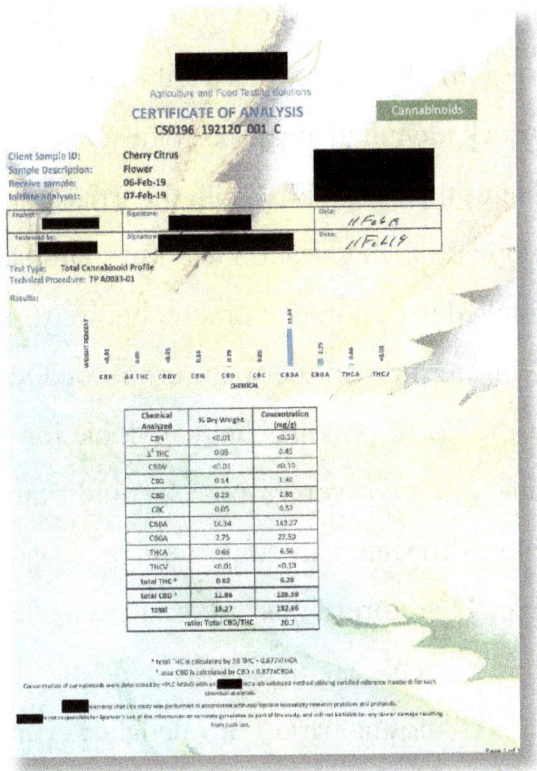

The certificate of analysis (COA) for our Cherry Citrus variety of industrial hemp showing the quantity of various cannabinoids, terpenes, and contaminants in the plant genetics.

Our farm manual contained the delegation of daily responsibilities among the various cultivation team members, including a daily irrigation system check and daily field cultivation check. The field cultivation checks identified plant performance in the areas of new growth, flower formation, pest infestation, and overall plant viability. These observations would eventually become the data collected and reported in this chapter as cultivation results. The irrigation check identified any leaks, blow outs, dry areas, or other problems that might arise with the irrigation system.

No one on the cultivation team was familiar with the recently installed irrigation equipment, but we soon became accustomed to turning on the water at the well and quickly moving to the corresponding field to look for leaks that might spring up. Our young plants could stand neither drought nor drowning in this tender stage of their development. Therefore, it was important to make repairs to the irrigation system as quickly as any problems were discovered or risk losing part of our fledgling crop.

During that first week, a number of university administrators, professors, and FAMU|CAFS staff came out to the farm to make observations of the grow. We were excited to have them visit, and they were equally delighted with the overall set-up of the research areas. The seedlings were well documented in our farm manual, which included copies of all required regulatory paperwork. The manual also included a field map drawn from the row markers and plant flags accounting for every seedling on our farm.

Through these actions, we began invalidating the litany of reasons used to keep industrial hemp away from small farms and black farmers in Florida.

This photo was taken from Field No. 1 on October 8.

The mistaken belief that hemp genetics acquisition and transportation of hemp were an insurmountable challenge to small farmers had been dispelled by the thousands of hemp seedlings we successfully transported onto our farm and transplanted into our soil. The fear mongering concern for security breaches and the confusion fomented around THC testing and compliance would be equally refuted by our actions during the hemp grow. We had proudly passed our first field inspection from FAMU|CAFS and were well underway on our hemp farming journey.

This photo was taken from Field No. 1 on October 11 showing the rapid burst of growth in the plants from the "stretching phase" that occurred over the brief three-day period between both photos.

Article III: Week Two Results

Data Recorded Oct. 11, 2019		
Avg. Temperature	High 88°F	Low 63°F
Avg. Humidity	69%	
Avg. Plant Height	Field No. 1 13¾"	Field No. 2 14.82"
Avg. Plant Width	Field No. 1 13¾"	Field No. 2 15.88"
Survival Percentage	97.54% survivability *30 total plants have died for undetermined reasons.	
Lunar Projection	Waxing in a gibbous phase at 95 percent illumination, with a full moon expected on the evening of October 13.	
Weed Pests	Nut grass has spread even further in larger patches between the rows and around the plants in both fields.	

The second week of the flowering phase is considered the "pistil development" stage. A pistil is the female reproductive organ of the hemp plant. If a feminized hemp plant goes unpollinated, these reproductive organs will begin flowering and develop buds. If, however, the pistil meets pollen, then the reproductive organs will put their energy into producing seeds, not buds.

Our plants showed emerging white pistils during the second week post-transplant, with new leaf formations continuing to appear on stems and lateral branches. The development of white pistils from the middle of the leaf bunches on the plant is a good sign of new flower production and the start of flowering. The small, wispy, white pistil hairs will develop and become buds at those locations where the big, broad leaves meet the main stem.

At this point, our industry partner suggested we develop a plan for fertilizing the crop in the upcoming weeks. A cannabis plant undergoing the flowering phase requires certain nutrients, or fertilizer, to produce maximum yields. We had two options for fertilizing our crop. The first option was the slow and tedious process of fertilizing the plants by hand. This would be brutal to accomplish but had been our method of choice for the Herculean excavation of the thousands of plant holes.

With the irrigation system, we had a second option: fertigation. As mentioned earlier, our industry partner wanted us to fertigate our hemp crop using a PVC chemical injection system that we could build ourselves or an automated commercial nutrient delivery system which we could purchase from one of their nursery vendors. The fertigation system would have to be connected directly to the residential water well in order to work properly.

We were worried about connecting any type of fertilizer or fertigation system to an active residential water well due to the risk of backflow and cross contamination. The residential well that watered Field Nos. 1 and 2 was owned and still actively used by Aunt Virginia, who lived near Field No. 1. My aunt had graciously agreed to share her residential water well with the hemp plants growing in Field Nos. 1 and 2. We weren't going to risk contaminating her drinking water with fertilizer chemicals.

F1-16

Week 2 – Field No. 1, Plant No. 16: Top view of white pistil development appearing in the middle of leaf bunches.

I contacted irrigationists and plumbers for suggestions regarding large-scale backflow preventors, check valves, or other devices that might protect the potable water supply at the residential water well. Not one could provide equipment or a design that would keep my aunt's water supply sufficiently free from possible backflow contamination caused by connecting it to a fertigation system. None of my family members participating in the pilot project had contemplated risking their health or life due to water contamination caused by fertilizing hemp. This was not the moment to take that risk.

Since there was no way to safely connect the fertigation system to the current water supply without risking cross-contamination, we decided not to use a fertigation system for the 2019 hemp grow. On October 10, 2019, we applied one-half teaspoon (½ tsp.) of an 18-12-6 fertilizer mixture, by hand, around the base of each plant in Field Nos. 1 and 2. The irrigation system was turned on for one hour in each field afterward.

Article IV: Pest Observation - October 12, 2019

Pests such as moles, rabbits, and deer were considered as possible threats to our newly planted crop. Expensive mitigation measures like high fencing, sound emitters, or light-flashing beacons were appraised, but we decided to go without any of them. Despite deer tracks appearing all over our hemp fields and some of the drip tape being pulled to-and-fro, a row-by-row examination of our plants showed that none had been eaten or even disturbed.

Nevertheless, we were still concerned and did not want frequent visits from any pests. My mother along with her sister, Aunt Virginia, suggested we try using a watered-down hot sauce spray to keep the furry vermin away. We purchased a couple of three-gallon lawn and garden sprayers from the local hardware store and mixed a water and hot sauce concoction. The concoction was sprayed directly on the hemp plants, and in the days that followed, we monitored our fields for more deer tracks. Even though the tracks returned, our plants remained undisturbed. It appeared that the deer and other vermin were curious about the hemp plants but didn't nibble. Not even when the plants had been "drizzled" with our homemade hot sauce. I admit to being pleasantly confused by the outcome of this pest encounter.

Article V: Week Three Results

Data Recorded Oct. 18, 2019		
Avg. Temperature	High 66°F	Low 61°F
Avg. Humidity	74%	
Avg. Plant Height	Field No. 1 14.6"	Field No. 2 15.36"
Avg. Plant Width	Field No. 1 16¼"	Field No. 2 16.83"
Survival Percentage	97.38% survivability *32 total plants have died for undetermined reasons.	
Lunar Projection	Waning gibbous phase at 80 percent illumination, with a third-quarter moon expected on October 21.	
Weed Pests	Nut grass continues to spread around the plants in both fields.	

The third week of the flowering phase is known as the "bud development" stage. As the stretch of the first few weeks slows down, the plant begins to form small buds that have white pistil hairs sticking straight out of them. These pistils signal where the flower buds will later appear.

The main colas, or central flower cluster, of the plants in each field continued to grow with multiple nodes appearing along the primary stem and some of the lateral branches. The cola forms along the upper portion of the main stems and large branches of a mature female cannabis plant. At the locations on the plant where pistils had formed, there now appeared the first signs of bud development.

Week 2 - Field No. 1, Plant No. 12: Side view.

During this week of being in the soil, we expected our plants would begin to show whether they were male or female. Female plants show signs of bud development, while male plants, on the other hand, show signs of seed growth. This is how we determined whether we had any male plants that needed to be removed. Fortunately, we didn't find any male plants in either Field No. 1 or Field No. 2.

Week 3 - Field No. 1, Plant No. 12: Side view showing bud development.

A male hemp plant growing seeds in place of flowers.

Article VI: Week Four Results

Data Recorded Oct. 25, 2019		
Avg. Temperature	High 86°F	Low 72°F
Avg. Humidity	75%	
Avg. Plant Height	Field No. 1 15.20"	Field No. 2 14.7"
Avg. Plant Width	Field No. 1 16.77"	Field No. 2 16.94"
Survival Percentage	97.38% survivability *32 total plants have died for undetermined reasons.	
Lunar Projection	Waning crescent phase at 10 percent illumination, with a new moon expected to occur on the evening of October 27.	
Weed Pests	Nut grass continues to spread around the plants in both fields.	
THC Compliance & Potency Analysis	Total THC 0.110%	Total CBD 2.79%

The fourth week of the flowering phase is known as the "larger buds" stage. The overall growth of the plants in Field Nos. 1 and 2 appeared to have slowed. However, there were more and larger buds growing on the main and lateral stems. The main colas of the plants in each field continued to grow, and the hairs of the pistils stuck out further. In addition, the plants produced a noticeable odor.

Week 3 – Field No. 1, Plant No. 12 side view.

We received an obligatory, non-regulatory farm visit from **FDACS** staff during that week. Being one of the first hemp farms meant the state might request entry on our premises for observational purposes; more so because we were the closest and only nearby site available. Despite this understanding, I was on pins and needles. Visions of "state regulators" coming to our fields and snatching hemp plants from their earthen beds invaded my every thought. I scantly slept in the days preceding that FDACS visit.

Week 4 – Field No. 1, Plant No. 12 side view showing larger bud development.

When the visit occurred, I was pleasantly surprised to find that the FDACS staff could not have been kinder or more professional to us as novice hemp farmers. They came out, clipped some plant samples, took several photos, and tried out their field testing and analysis procedures. As a cannabis advocate afflicted with Green Rush fever, it was uplifting to assist a group of state regulators in their determination of industry standards and criteria for future enforcement activity. They even thanked us for allowing them to come out after finishing their work. Overall, it was a satisfying experience.

A few days after the FDACS staff visit, we prepared to clip our first hemp samples for submission to a private, third-party lab for THC compliance and potency analysis. To be THC compliant by the time we reached harvest, our crop had to be at or below the legal THC limit of 0.3 percent, with only a small margin of error. Because of the selected Cherry Citrus genetics, we expected our plants to maintain their THC compliance through maturity. The hemp flower sample measured four to six inches in length and was clipped and dried from a previously, randomly selected hemp plant in Field Nos. 1 and 2.

Article VII: Week Five Results

Data Recorded Nov. 1, 2019		
Avg. Temperature	High 64°F	Low 41°F
Avg. Humidity	72%	
Avg. Plant Height	Field No. 1 14.64"	Field No. 2 15.71"
Avg. Plant Width	Field No. 1 17.03"	Field No. 2 16.95"
Survival Percentage	97.38% survivability *32 total plants have died for undetermined reasons.	
Lunar Projection	Waxing in a crescent phase at 22 percent illumination, with a first-quarter moon expected to occur on the evening of November 4.	
Weed Pests	Nut grass continues to spread around the plants in both fields. Suppression efforts undertaken using non-chemical method of gas-powered weed trimmer.	
THC Compliance & Potency Analysis	Total THC 0.345%	Total CBD 8.76%

In the fifth week of the flowering phase, the trichomes were in the early stage of development and appeared as clear hairs or beads of dew on the developing buds and broad leaves. Trichomes protect the hemp plant from external threats and help it to survive extreme conditions. They are the resin glands of the hemp plant containing THC, CBD, terpenes, and other active cannabinoids. These compounds provide the plant with protection from UV rays and predators. The appearance of these tiny hair-like growths is an indication of healthy and potent hemp plants.

Week 5 – Field No. 1, Plant No. 12: top view.

Overall, our plants appeared fatter as the buds grew thicker. The main colas had become denser and better developed. What appeared to be new bud formations were growing on the lateral stems along the main cola. The hairs of the pistils continued to get longer on the top of the plants. Some of the white pistil hairs appeared to turn darker and curl toward the bud formations. The odor of the plants intensified becoming more earthen in aroma.

141

Week 5 - Trichomes appear as clear hairs or beads of dew on the buds and broad leaves.

Week 5 – Field No. 1, Plant No. 12: side view.

Article VIII: Pest Observations - November 4, 2019

On November 4, we hosted a farm visit with FAMU|CAFS professor Dr. Lambert Kanga and his research group. Dr. Kanga came out, at our request, to observe red ants Cousin George Jr. had first noticed around the hemp plants during their second week of flowering. Dr. Kanga confirmed the possibility that the deaths of a number of our plants were due to red ant infestations at the roots and along those plant's stems.

Dr. Kanga advised us to undertake spot treatments to eliminate the immediate threat of the red ants in the field. Two ant infested plants that were still alive in Field No. 1 and two ant infested plants that were still alive in Field No. 2 were treated with boiling water poured directly at the base of the stems and along any branches containing fire ants. We then removed all dead infested plants at their root plugs and poured boiling hot water over the area vacated by the root plugs. Further observation of the removed dead plants revealed poor root formation along with a few fire ants still attached around the stem base. There weren't any signs of ant chewing patterns on plant leaves or stalks.

Dr. Kanga followed up the farm visit with a written observation detailing several other pests discovered on and around our hemp plants, including mites, whiteflies, and aphids. His observation also revealed a Septoria fungus infection on several of our plants. Although a Septoria fungus infection is not necessarily fatal to plants, when left untreated, it rapidly spreads on the plant leaves and can quickly weaken plants making them unable to bear fruit to maturity. In terms of our industrial hemp, that fruit would be our flower production, which we could not afford to diminish any further this grow season.

Dr. Kanga explained that an untreated Septoria infection can cause leaves to spot, turn yellow, dry out, and fall off. Such defoliation would weaken the plant and send it into quick decline. Dr. Kanga further advised that the options for treating Septoria infection include immediate removal of infected leaves by hand-pruning with shears, application of an organic fungicide containing copper or potassium bicarbonate, or application of a low-toxicity chemical fungicide like chlorothalonil (e.g., Fungonil and Daconil). We chose to prune as many of the infected plants as we could by hand. Ultimately, we were unable to prune away all of the infection spots on the crop. This challenge demonstrated how useful it would have been to build and connect a fertigation system to our irrigation system. Again, I quietly resolved to myself that our next grow would be different.

The area circled in red shows an infestation of red ants along the primary and lateral stems of a hemp plant.

The area circled in red shows an infestation of red ants along the primary and lateral stems of a hemp plant.

Article IX: Week Six Results

Data Recorded Nov. 8, 2019		
Avg. Temperature	High 66°F	Low 63°F
Avg. Humidity	84%	
Avg. Plant Height	Field No. 1 15.19"	Field No. 2 16"
Avg. Plant Width	Field No. 1 16.57"	Field No. 2 16.51"
Survival Percentage	97.38% survivability *32 total plants have died for undetermined reasons.	
Lunar Projection	Waxing in a gibbous phase at 85 percent illumination, with a full moon expected to occur on the evening of November 12.	
Weed Pests	Growth of nut grass mitigated and suppressed by removal with a gas-powered weed trimmer.	
THC Compliance & Potency Analysis	Total THC 0.278%	Total CBD 7.07%

In the sixth week of the flowering phase, we started to see the effects from the fertilizer application of October 29. Although there were no measurable height or width increases in the plants, their main colas and broad leaf formations turned a darker green color. More and thicker buds had formed on the main colas and lateral branches, and some pistil hairs appeared to have turned a reddish-purple color.

Week 6 - Translucent trichomes in their early stage of development producing cannabinoids.

The trichomes observed on the buds and broad leaves were still in their early development and appeared as clear hairs or beads of clear dew. The translucent coloring of these trichomes indicated they were still producing cannabinoids. The plants had a pungent odor and their leaves felt sticky to the touch from resin.

During this week, I attended a live broadcast of the podcast, REGULATED. REGULATED covers national and state regulatory law from the perspective of two regulatory lawyers. One of those lawyers is Mr. Christian Bax who served as the director of Florida's Office of Medical Marijuana Use from 2015 through 2018. The other lawyer is Mr. Tony Glover, who served as director of the Division of Pari-mutuel Wagering and deputy director of the Division of Alcoholic Beverages and Tobacco.

Mr. Tony Glover (left) and Mr. Christian Bax (right), are the regulatory lawyers who broadcast the REGULATED podcast.

This particular episode of the podcast featured a discussion about Florida's regulated cannabis and hemp industries. Mr. Bax, being the regulatory lawyer most attuned to the cannabis industry, was the primary speaker on the topic of hemp. Although he had a bleak forecast for smaller-scale hemp farming in Florida, Mr. Bax embraced the belief he shared with Mr. Glover that social justice measures were necessary in the cannabis industry, specifically on behalf of black people.

They discussed measures to combat the lack of inclusion and business ownership in the cannabis industry, as well as measures to stop racial profiling in controlled-substance law enforcement. Both lawyers noted that the legalization of industrial hemp was already having a positive effect on social justice. Prosecutors nationwide were suspending prosecution of certain cannabis-related offenses where industrial hemp was now legal to grow, because legal industrial hemp is virtually indistinguishable from illegal marijuana. I was gratified to hear that the social justice sentiments espoused by both podcast lawyers were similar to my own. The overall discussion was lively and well received by the audience of students and faculty members. I was glad to have attended.

Article X: Week Seven Results

Data Recorded Nov. 15, 2019		
Avg. Temperature	High 54°F	Low 46°F
Avg. Humidity	92%	
Avg. Plant Height	Field No. 1 14.61"	Field No. 2 15.76"
Avg. Plant Width	Field No. 1 16.84"	Field No. 2 18.78"
Survival Percentage	97.38% survivability *32 total plants have died for undetermined reasons.	
Lunar Projection	Waning gibbous phase with 91 percent illumination, and a third-quarter moon was expected on the evening of November 19.	
Weed Pests	Continued mitigation and suppression of nut grass growth using gas-powered weed trimmer.	
THC Compliance & Potency Analysis	Total THC 0.342%	Total CBD 8.58%

In the seventh week of the flowering phase, there was no measurable height or width increase in the plants. The main colas appeared to have stopped adding new buds or getting thicker. The broad leaf formations turned an even darker evergreen, and the stems connected to the broad leaves were now purple, matching the color of the broad leaves. Existing pistil hairs continued to turn darker and curl toward the stem, forming buds with new white pistil hairs still developing. The trichomes were still translucent, but some cloudiness was beginning to appear. The plants still had their pungent smell.

On November 17, 2019, we applied one teaspoon (1 tsp.) of a 4-35-26 fertilizer mixture by hand around the base of the soil for each plant in Field Nos. 1 and 2. The irrigation system was turned on for one hour in each field after application of the fertilizer.

Week 7 – Field No. 1, Plant No. 12: Side view.

Week 7 – Field No. 1, Plant No. 12: Top view.

Article XI: Week Eight Results

Data Recorded Nov. 22, 2019		
Avg. Temperature	High 73°F	Low 45°F
Avg. Humidity	84%	
Avg. Plant Height	Field No. 1 14.61"	Field No. 2 15.76"
Avg. Plant Width	Field No. 1 16.84"	Field No. 2 18.78"
Survival Percentage	97.38% survivability *32 total plants have died for undetermined reasons.	
Lunar Projection	Waning crescent phase with 22 percent illumination, going to a new moon phase on November 26.	
Weed Pests	Continued mitigation and suppression of nut grass growth using gas-powered weed trimmer.	
THC Compliance & Potency Analysis	Total THC 0.334%	Total CBD 9.09%

In the eighth week of the flowering phase, the new growth broad leaves were dark purple. However, the lower broad leaves showed signs of yellowing because the plant's upper canopy shaded them from sunlight. The existing buds had become stiffer, thicker, and greener with frosty amber pistils curled around them. The translucence had left the trichomes, and they had become cloudy in color, indicating that the plants were ready to harvest. Overall, the plants were more fragrant.

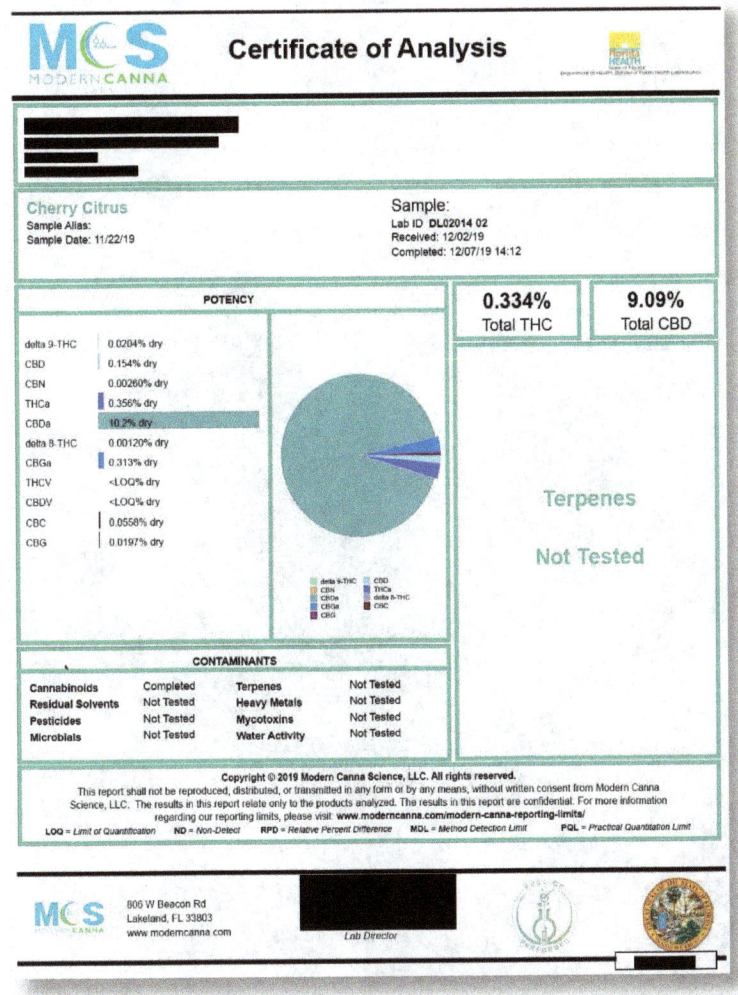

Week 8 - Cherry Citrus variety lab report showing THC compliance and CBD potency analysis.

Week 8 – Field No. 1, Plant No. 12: Side view.

Week 8 - The trichomes have become cloudy in color indicating that the plant is ready to harvest.

Week 8 – Field No. 1, Plant No. 12: Top view.

Article XII: Week Nine Results

Data Recorded Nov. 29, 2019		
Avg. Temperature	High 77°F	Low 46°F
Avg. Humidity	56%	
Avg. Plant Height	Field No. 1 14.61"	Field No. 2 15.76"
Avg. Plant Width	Field No. 1 16.84"	Field No. 2 18.78"
Survival Percentage	97.38% survivability *32 total plants have died for undetermined reasons.	
Lunar Projection	Waxing crescent phase with 10 percent illumination, going to a first quarter (or half-moon) phase on December 4.	
Weed Pests	Continued mitigation and suppression of nut grass growth using gas-powered weed trimmer.	
THC Compliance & Potency Analysis	No sample submitted.	No sample submitted.

In the ninth week of the flowering phase, 12 days after the last application of fertilization, new growth broad leaves turned from a purple color to a darker maroon. Cloudy trichomes covered the leaves and buds. Buds continued to stiffen and thicken as reddish-purple pistil hairs curled around them. The plants still have a strong pungent odor. Since we were less than a week away from harvest, pursuant to Rule 5B-57.013 we no longer needed to send flower samples for lab analysis. The test results for week eight would be the last recorded before our harvest.

Week 9 – A plant in Field No. 1 before harvest.

Week 9 - A cola full of buds ready for harvest in Field No. 1

Chapter 18: Harvesting and Drying

Article I: Preparations

The general principles for properly harvesting and drying cannabis required knowing the atmospheric conditions where we intended to dry the crop. Those atmospheric conditions would be the opposition against which our drying system would contend. This information would serve as the basis for properly outfitting our drying room with the right equipment to promote the removal of moisture from our industrial hemp plants.

In the right atmospheric conditions, a farmer might only need to hang their hemp harvest from horizontal racks in a well-ventilated area to dry them. That would be similar to how tobacco leaves are sun dried and cured for cigars in Caribbean climates. However, planting our hemp in North

Florida in late September for harvest in December put us in the category of winter seasonal crops; far from the Caribbean tropical climate.

Typically, the northern growing region of Florida is known for winter temperatures that can drop to below freezing at least once a year, along with the chance of snow. Winter in north Florida tends to be dry, with an average daily temperature of 39°F and average dew points of 40 to 50 without a lot of rainfall. Colder temperatures cause dryness by reducing humidity; that reduction being the result of colder air holding less water vapor than warmer air.

Unlike Caribbean cigar tobacco growers who have milder climates more conducive to outdoor crop drying, our North Florida winter temperatures would not provide a good outdoor atmosphere in which to air dry our harvested industrial hemp. We would need an indoor location for drying. As a result, the space we chose to dry our harvested hemp was inside the vacant farmhouse/office of Cousin George Jr.'s father. We had briefly reviewed the space with our industry partner during their earlier farm visit. It was now time for a more detailed review of the space.

Cousin George Jr. had enclosed the 1,000 square foot carport space with cinder blocks. The carport's ten-foot ceiling was comprised of plywood panels, which also served as the floor for the first story. In preparing this space as a drying room, we first removed all debris and contaminates. In order to remove any mold, mildew, or other bacterial matter that might adversely affect the plants, we cleaned and sanitized the floors, walls, and overhead rafters with liquid mold remover and bleach water. To help keep moisture out, we used plastic sheeting and duct tape to seal the plywood board junctions on the first-story side of the plywood ceiling panels.

We then gathered the specialized equipment our industry partners advised we needed to achieve the specifications for optimal drying. The specifications for optimal drying that the drying room needed to maintain were a temperature between 60°F to 70°F, with a humidity level of around 50 percent. The array of equipment needed to exert this level of climate control in the uninsulated, enclosed carport included pedestal fans, box fans, trellis netting, dehumidifiers, space heaters, digital hygrometers, and digital moisture meters.

165

Drying and curing room with trellis netting.

Drying and curing room set up with dehumidifiers and netting.

Cousin George Jr. further suggested we install green lighting throughout the drying room to continue our basement work in a darkened space. He said this would avoid putting any direct, bright light on the drying flowers, which might further degrade the cannabinoids within them. We quickly implemented his suggestion, making the space feel like it had some nefarious, hole-in-the-wall agenda. It looked wicked and seemed to work just fine.

We continued our preparations by attaching trellis netting to the rafters and strategically setting up space heaters and fans around the room. The fans were used to circulate air throughout the room. Even though space heaters don't actually change the overall humidity of a room, they can be utilized to change the relative humidity of a drying space. The relative humidity measures how much water vapor is in the air relative to the temperature.

The space heaters help remove moisture from the plant, because warm air picks up and holds much more water than cold air. Consequently, warmer air would carry more moisture away from the hemp plants. That moist, warm air would still need to be removed from the atmosphere of the room to avoid any further impediment to continuing the drying process. That task would be handled by the dehumidifiers.

The dehumidifiers recommended by our industry partner were large, commercial-scale units that could remove at least 250 gallons per minute (gpm) of moisture from the atmosphere of a 1,000 square foot drying room. Our industry partner cautioned us that the main problem with using large dehumidifiers is the risk of drying out the air in the room too fast, negatively affecting the valuable hemp buds.

A digital hygrometer with an acceptable reading
for our drying process to begin.

We would rely on our digital hygrometer and digital moisture meter to help maintain the proper climate conditions in the drying room. A hygrometer tracks the humidity and temperature in the drying room, and a moisture meter monitors the level of moisture in the harvested plant matter. These monitoring devices would prove their worth as our crop underwent the drying process.

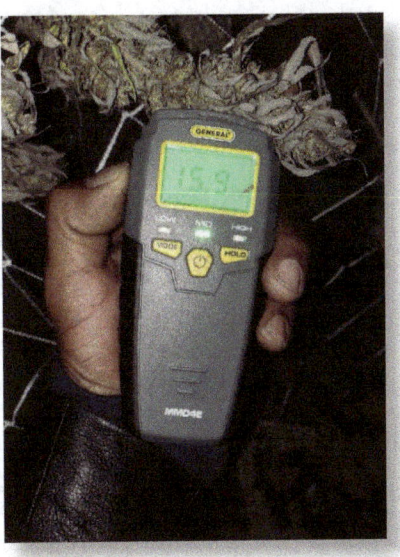

A digital moisture meter being used to measure the moisture content of drying plant flowers.

Harvesting involves gathering and collecting a crop to remove it from the field. We chose to follow a harvesting methodology designed to avoid exposing our plants to any extra moisture. We broke down our harvesting process into three main functions; reaping, trimming and bundling, and transport to the indoor drying location.

Reaping is the process of cutting and gathering a crop. Traditional garden shears would suffice to cut down our short hemp plants. We determined that it would be necessary to gather our crop during a weather interval optimal for the removal of moisture. The presence of dew, frost, or other moisture would only prolong the drying process. So, to promote drying we would avoid reaping early in the morning and during, or after, rain.

Next in the harvesting process came the tasks of trimming and bundling the plants. During trimming, the harvester has the opportunity to remove moisture from the plant by cutting away stems, broad leaves, leaves protruding from buds, and any other non-floral matter that naturally contains water. There are generally two recognized, standard methods for trimming cannabis plants; wet trimming and dry trimming. Wet trimming is when the non-floral matter on the plant is trimmed before the plant is dry. The wet trimming process has been hailed for decreasing the overall drying time needed for a cannabis plant because there is less wet plant matter to dry.

Dry trimming, on the other hand, involves reaping the entire plant and leaving the trimming of non-floral matter until after the plant has dried. We chose to dry trim our plants because of our concerns for inclement weather. Our goal was to get the crop out of the field and indoors as quickly as possible. Dry trimming would allow us to accomplish that.

After a group of plants was reaped, we would transport them to the drying room. Since we were not transporting the hemp across noncontiguous locations, the movement of the harvested crop would not require a transportation and movement plan pursuant to Rule 5B-57.013, F.A.C. Reaped plants would quickly be placed in large plastic bins and taken into the drying room.

Cultivation team members Mother Althea (left) and Aunt Virginia (right) harvesting plants in Field No. 1.

We broke down the drying process into four main functions; hanging, monitoring, trimming, and storing the dried plants. We would hang our harvested plants upside down to make the best use of the available vertical drying

pistils was also a good indicator that our hemp plants were ready to leave the soil. The plants did not appear damaged from the frost events, and we notified our industry partners and FAMU|CAFS representatives of our new intended harvest date. Again, I quietly resolved to myself that our next grow would be different.

On December 6, 2019, approximately ten weeks after transplant into our soil, we harvested our industrial hemp crop from Field Nos. 1 and 2. Our cultivation team easily transitioned into a harvest team after being supplied with the five-eighths-inch pruning shears, clean 30-gallon plastic bins, and specific instructions for dry trimming our plants before bundling them for transport. Pruning shears were used to hand cut each plant at the base of the primary stem approximately two to three inches above the soil. We did not measure the moisture content of the harvested hemp plants when they were reaped. We did not take note of any moisture readings until the indoor drying process had commenced.

Harvest team members who had participated in planting day were surprised to find the hemp plants were covered with sticky resin, making the reaping task slower and more methodical. Harvest team members also remarked about the plants highly pungent odor. However, I had lost my sense of smell and taste two weeks earlier from a bad flu-like illness that I couldn't shake and was unable to smell even a waft of our harvest's lovely cannabis aroma.

Hemp plants hanging in drying room.

The research we gathered advised that a properly dried hemp bud should be squeezable, like a plump marshmallow; not too stiff or dry. The research also warned that the drying process for cannabis plants must be slow and carefully monitored. Rushing the process, over-drying the plants, or inconsistently drying them would result in dried-out, powdery, or otherwise fragile buds with diminished levels of cannabinoids and terpenes. Diminished levels of cannabinoids and terpenes meant a lesser quality, lower valued bud.

Worse would be an incomplete drying process that allowed moisture to remain in the plant product, thereby increasing the risk of mold and mildew contamination to the end user. Such unseen contamination inside of a bud could ultimately wipe out an entire crop that was stored with the compromised flower. This was an outcome to avoid at all costs.

Reaped plants were bundled into large thirty-gallon plastic bins, put on trailers attached to four-wheel ATVs, and taken to the drying room. The harvested plants were hung upside down in the trellis netting to dry, leaving enough space between each plant to allow sufficient air flow. We had chosen to dry trim our plants, which left the removal of non-floral matter until after drying was complete. When the dried flowers and leaves reached the targeted moisture reading, the remaining stems, broad leaves, and any other non-floral matter that naturally contained water were trimmed away and removed. We continued the drying process by placing the dried hemp flower in large brown paper bags and keeping those paper bags in the dark until the buds could be vacuum sealed for long-term storage.

Hanging hemp plants.

Dried plant flowers and leaves removed by hand from their stalks and branches.

Dried plant flowers and leaves.

Chapter 19: Agricultural Marketing and Commercialization

Article I: The Initial Goal

Our harvest goals for Field Nos. 1 and 2 were to grow hemp for flower production that would be used for CBD oil extraction. With the crop successfully harvested, our attention turned to the market value of our hemp flower. The value would be determined by its quality, as measured by lab analysis, and its dry weight, as measured in pounds or tons.

Our industry partner specified that, for their chosen hemp processing vendor to extract oil from our harvest, we would need at least 100 pounds of dried floral matter that had a maximum water content level between ten percent and 12 percent. They explained that the CBD oil extraction process has its own universe of factors to consider for efficient operation. These factors were different and apart from the cultivation and drying issues we had faced up to that point as farmers. The CBD oil extractors do not run the hemp buds through a crop-pressing machine to extract the oil by boiling the resulting liquid in a cast iron kettle. It would have been really convenient for our family farm if the CBD oil extraction process worked that way, but it did not.

CBD oil extraction was a much more scientifically based process that required calibrating equipment with uniform crop specifications. I will not digress to further explain that here, except to say, typically, the oil extraction processors need a minimum weight of hemp flower material of a specific quality, cut, and moisture content in order to operate their extraction equipment. In fact, certain CBD oil extraction equipment that has been designed for industrial and agricultural processing demands will only properly function with loads of more than 3,000 pounds of hemp per run.

None of us on the family farm knew specifically how much oil could be extracted from our harvested flower. Even though we easily met the oil extractor's moisture content requirement, our 18.5 pounds of harvested flower hardly satisfied the weight requirement for an efficient oil extraction process. We were well below the minimum threshold and would need more hemp flower than we had harvested in order to ensure that oil extraction did not cost us more money than the end product we would obtain. Again, the straightforward farming maxim proved prophetic. "We had to grow enough of the crop before we would have enough of it to decide what to do with it." Alternative options for commercializing our crop would need to be considered.

Clipped sample of dried plant flower.

Article II: The New Goal

Whether it was new cannabis varieties or new cannabis products, Cousin George Jr. knew the latest trends in cannabis and cannabis products. He was some thirty years younger than me and the executives of our industry partner and had performed market research in the City of Tallahassee throughout our grow season. Cousin George Jr. had discovered that the Green Rush had quietly found its way to Tallahassee smoke shops which were now selling smokeable hemp flower by the bud. We were excited imagining the potential local market opportunity, since Tallahassee was home to more than 70,000 college-aged people, some of whom were happily buying smokeable hemp flower.

Cousin George Jr. implored us to visit local cannabis shops and confirm the facts through our own market research. I researched shops throughout Tallahassee and as far away as Quincy, Florida. The executives of our industry partner visited stores in the Central Florida region located around their nursery. We were all delighted to discover that my cousin was right.

Store vendors were taking advantage of the changes in state law regarding hemp and were selling raw hemp flower to the public. Moreover, the stores we contacted agreed to sell our harvested hemp if it was THC compliant and grown on a properly licensed farm under state law. Selling hemp flower directly to local stores had not been a market use option when we'd first put our seedlings in the ground. I needed to quickly perform further research to determine the viability of this smokeable hemp option for our family farm.

Market research concerning smokeable hemp became necessary.

I discovered that smokeable hemp flower provided the same therapeutic advantages found in its cannabidiol (CBD) content without an inebriating high. In addition, smokeable hemp had not yet been banned in Florida as a legal method of ingesting industrial hemp. In 2018, an estimated $70.6 million of CBD pre-roll and raw flower were sold as smokeable hemp in the United States. Indicative of this market's strength was the fact that smokeable hemp flower sold for between $150.00 and $350.00 a pound throughout most of that year.

Taking harvested hemp flower directly from field to shelf would eliminate the costs involved in extracting or machine processing the hemp flower into oil, making the crop that much more valuable as an agricultural commodity for small farms and black farmers. The twist of fate that turned our harvest away from oil extraction might have been the type of break we needed to claw back some financial success for our small farm venture.

Article III: The Ethical Dilemma

Florida hemp law at that time provided assurances to the end user of a harvested hemp crop by requiring that compliance and potency tests be performed 30 days before harvest. However, as cottage industry farmers and sugar cane syrup makers, our family understood that producing a quality agricultural product also meant taking responsibility to ensure that it was safe for public consumption. At the time we were considering the smokeable market use option for our hemp harvest, there was no Florida statute requiring that a licensed hemp farmer test their harvest for toxins, contaminants, or safeness for human consumption.

Florida hemp law defined contaminants unsafe for human consumption as "...any microbe, fungus, yeast, mildew, herbicide, pesticide, fungicide, residual solvent, metal, or other contaminant found in any amount that exceeds any of the accepted limitations..." as established by state law. Nevertheless, this directive regarding contaminants unsafe for human consumption was only applicable to "hemp extract" distributed and sold in the state. Hemp extract, at that time, was defined as "...a substance or compound intended for ingestion that is derived from or

185

contains hemp and that does not contain other controlled substances."

By that language, the definition did not include the *Cannabis sativa L.* plant or any part of the plant. Notwithstanding the question of whether the term "consumption," as used in F.S. 581.217(7)(a)(3), included "inhalation" and despite the lack of regulatory guidance from the state in this area, we decided it best to test our dried hemp flower for its purity which would indicate its safety for human consumption.

The testing was accomplished by performing a full panel of biological contamination tests, including screening for the presence of chemical contaminants (e.g., residual pesticides, residual solvents), microbial contaminants (e.g., bacteria, mycotoxins), physical contaminants (e.g., dust, dirt, hair), and heavy metals (e.g., arsenic, cadmium, mercury, lead). This type of testing is important because industrial hemp is considered a bio-accumulator that will absorb those types of contaminants if the plant's root system encounters them.

We acquired the ethical responsibility to further test our crop as a result of changing our harvest goal to the sale of smokable, dried hemp flower. Before selling our raw dried flower crop to the public, we had to know that it was safe for human consumption through inhalation. The other interested family members quickly gave their consent to submitting a sample dried flower from the harvest to a third-party cannabis testing lab for a full panel of biological contamination testing.

The results of the sample were good, revealing an absence of microbials, including yeast and mold. Nor were any of the various, listed pesticides detected in our dried flower sample. Mycotoxins appeared within acceptable levels. The heavy metals portion of the test further showed arsenic, cadmium, and mercury were present, but well within acceptable amounts. However, the level of lead appearing in our sample was over the prescribed limit.

We were surprised to find that our dried hemp flower contained any lead at all. The results of our pre-planting soil analysis did not forewarn of such, nor did our cultivation methods involve or infuse any heavy metals into the soil. Notwithstanding how it got there, we deemed the heightened level of lead in our dried flower as unhealthy and unacceptable. Despite our dried hemp flower being compliant with all state laws for potency, we deemed it tainted. The expenses of the grow were huge and could not be recouped without revenue from this harvest. Taking that kind of a financial loss at this early juncture of our hemp farming venture might be the irreparable event from which we would not recover.

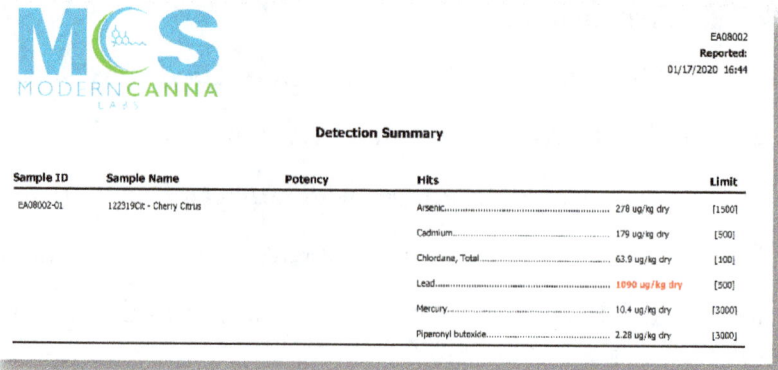

Results of biological contamination panel test
showing presence of heavy metals.

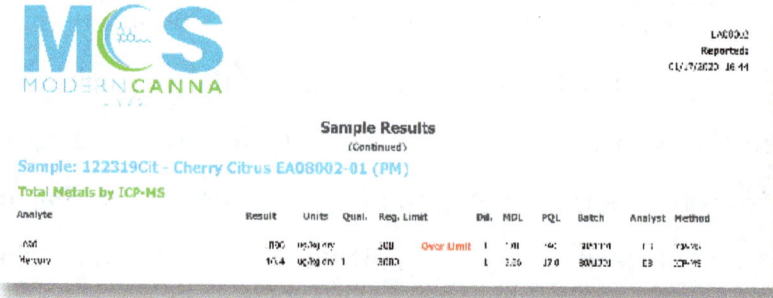

Page two of results from panel test.

This dilemma made me reflect on the times when Granddaddy Sam had driven his truck down from our Red Hills farmland into the City of Tallahassee, to sell his farm produce to friends and neighbors throughout the community. He provided them with more than what a financial bond or liability policy could offer. His accountability and responsibility to the community were based on the food he grew on his own land and sold from the back of his truck. As a small farmer, he breathed, ate, and lived locally.

Granddaddy Sam would consider it an open shame if he knowingly sold bad or tainted produce. There was no way we were going to turn against that. Our family's farming heritage and cottage industry background compelled us to decide that we could not ethically sell the tainted dried hemp flower. Despite the decision not to sell our hemp, we were still curious to know the estimated market value of our crop. I searched local and regional news, trying to determine the prevailing market price for industrial hemp of potency similar to ours.

During my investigation, I came across a news story about the three-day International Hemp Auction and

Market held in late November 2019 in Franklin, Tennessee. This would be the first auction of its kind. However, there were more hemp growers in attendance at the auction to sell their product than buyers willing to pay for it. Prices were driven down leaving smokeable hemp flower priced as low as $75.00 per pound due to the oversupply. At $75.00 per pound, our 18.8 pounds of flower would have earned us a paltry $1,387.50, which would not have even covered the purchase of our seedlings.

As I recuperated in January 2020 from the flu-like illness that had plagued me since the previous December, I considered the disheartening news about the downturn in hemp pricing. I sent an email to the other members of my family who were still interested in hemp farming. That email contained a foreboding outlook for the future of our hemp farming endeavor.

I remarked that the out-of-state hemp already on Tallahassee store shelves had created a supply glut. This was caused by farmers nationally outgrowing the capacity to process or sell industrial hemp in their own regions. I further explained that the oversupply of harvested hemp had pushed all market prices way down for hemp farmers. In fact, the hemp we planted on September 27 lost value in the ground each week when other, out-of-state hemp crops were being harvested. Our hopeful estimate for market prices when we planted was up to $300.00 per pound for hemp flower. However, the reality a few months later in November 2019 was that the oversupply had pushed the price down to as little as $100.00 to $150.00 per pound.

I concluded my email by reluctantly acknowledging that we needed to suspend our hemp growing activities until both the lead issue resulting from the plant's bioaccumulation capability was resolved, and market prices were better suited for us to make a profit. Given current market conditions, investing in a Spring 2020 grow would be akin to a vanity project not based on sound business principles.

Part IV: Concluding Observations

Chapter 20: In Summary

Despite the late issuance of the planting permit and the challenges that seemed to beset our small hemp grow, we still managed to cultivate our hemp plants to maturity. Honestly, just before harvest, our crop looked like a field full of sad little Charlie Brown Christmas trees. However, instead of their pitiful stems and thin branches sagging with tree ornaments, our plants' little limbs hung low, laden with thick, sticky buds. Aside from the meager appearance of the crop, our small, black-owned, family farm considered the THC compliance, survival, and flowering of these little hemp plants to be a huge win.

Our successful harvest invalidated several false platitudes that had been misguidedly proffered to keep industrial hemp away from small farms and black-owned farms in Florida. We obliterated the misbelief that the acquisition and transportation of hemp genetics would be an insurmountable challenge small farms could not solve. The fear-mongering concerns for security breaches and thefts were likewise disproven by the thousands of plants we managed to successfully cultivate to harvest unmolested. Finally, the chiding reproach that small farmers would be confused by THC testing and compliance was entirely destroyed, as we successfully tested our crop for THC compliance, weekly, throughout the grow. There is nothing inherent about industrial hemp or its cultivation that prevents small farmers or black farmers from successfully growing it on their own farms.

Moreover, the industrial hemp industry could significantly enhance its status with the black community by supporting greater diversity among its farming participants. Black-owned hemp farms would help reduce some of the economic challenges and environmental racism currently facing many rural minority communities where these farms operate. Farming hemp on open croplands makes formerly fallow, unproductive land financially productive and less susceptible to acquisition by unethical farms that recklessly create negative environmental impacts in rural communities of color.

The black-owned hemp farms in those communities would create jobs and introduce a new revenue stream. A larger, more diverse group of hemp farmers would change the overall outlook for rural communities and promote overall growth of the industry. Increasing the number of black-owned hemp farmers might also result in improving the cannabis industry's access to the billions of dollars of spending power in black communities nationwide. The influx of more familiar faces from whom the consumer can choose to purchase their product is a powerful marketing factor.

In these ways, hemp farming by black farmers could forestall the negative impacts of food deserts, economic distress, and environmental racism that currently afflict many rural farming communities of color. Even though our venture did not achieve financial success, our spirits soared high with hope and jubilation from the small triumphs achieved along the journey. This was my Green Rush Fever story, and I end it by saying next time, our hemp grow will be different.

Thanks for reading.

If you enjoyed this book, please consider leaving an honest review on your favorite store or social media website.

Bibliography

Additional information concerning the local weather forecast was compiled from data collected at https://www.timeanddate.com and www.wunderground.com.

Information concerning lunar projections was compiled from data collected at https://www.timeanddate.com/moon/phases/ and https://www.moongiant.com/.

2 NCAC 62.0101 et seq. (Ch. 62- Industrial Hemp Commission) (2017).

Section 581.217, F.S. (State Hemp Program) and see proposed Rule 5B-57.014, F.A.C.

American Civil Liberties Union. (2021 Research Report). A Tale of Two Countries: Racially Targeted Arrests in The Era of Marijuana Reform. Retrieved from https://www.aclu.org/report/tale-two-countries-racially-targeted-arrests-era-marijuana-reform/.

Arnsdorf, Isaac. (June 26, 2019). How a Top Chicken Company Cut Off Black Farmers, One by One. *ProPublica.* Retrieved from https://www.propublica.org/article/how-a-top-chicken-company-cut-off-black-farmers-one-by-one/.

Barth, Brian. (July 9, 2018). So, You Want to Be a Hemp Farmer? *Modern Farmer Media.*

Browning, Pamela, and others. U.S. Commission on Civil Rights, Washington D.C. (Feb. 1982). The Decline of Black Farming in

America. *A Report of the U.S. Commission on Civil Rights.* Retrieved from https://files.eric.ed.gov/fulltext/ED222604.pdf.

C-SPAN. (April 23, 1997). Agricultural Department Loan Discrimination. Retrieved from https://www.c-span.org/video/?80632-1/agriculture-department-loan-discrimination/.

Clayton, Bob. (Oct. 2017). *Industrial Hemp is Five Crops.* Published by Florida Hemp Processing, LLC.

Draper, Robert. (April 24, 2012). *Do Not Ask What Good We Do: Inside the U.S. House of Representatives.*

Drotleff, Laura. (December 9, 2019). Rocky start for inaugural hemp auction, but more events are planned. *Hemp Industry Daily.*

Fine, Doug. (June 25, 2014). Op-Ed: A tip for American farmers: Grow hemp, make money. *Los Angeles Times.* Retrieved from https://www.latimes.com/opinion/op-ed/la-oe-fine-hemp-marijuana-legalize-20140626-story.html/.

Fornadel PhD, Andrew P., Davis, Daniel L., Clifford PhD, Robert H., and Kuzdzal PhD, Scott A. (Vol. 1, Issue 1, March 1, 2018). Metals in Cannabis and Related Substances—Regulations and Analytical Methodologies. *Cannabis Science and Technology.* Retrieved from https://www.cannabissciencetech.com/article/metals-cannabis-and-related-substances%E2%80%94regulations-and-analytical-methodologies/.

"Heritage." Merriam-Webster.com Dictionary, Merriam-Webster, https://www.merriam-webster.com/dictionary/heritage. Accessed 2019.

"Heritage." Collins English Dictionary - Complete & Unabridged 2012 Digital Edition, https://www.dictionary.com/browse/heritage. Accessed 2019.

Jagannathan, Meera. (Nov. 12, 2019). People of color are reclaiming their place in a cannabis industry 'built on the backs of people from marginalized communities'. *MarketWatch, Inc.* Retrieved from https://www.marketwatch.com/story/people-of-color-are-claiming-their-place-in-a-cannabis-industry-built-on-the-backs-of-people-from-marginalized-communities-2019-08-05/.

Johnson, Renee, Specialist in Agricultural Policy, and Monke, Jim, Specialist in Agricultural Policy. (March 8, 2019). 2018 Farm Bill Primer: What Is the Farm Bill? *Congressional Research Service Report IF11126, Ver. 2.* Retrieved from https://crsreports.congress.gov/product/pdf/IF/IF11126/.

Leafly Staff. (April 1, 2020). CBD vs. THC: What's the difference? *Leafly.* Retrieved from https://www.leafly.com/news/cbd/cbd-vs-thc/.

"Legacy." Merriam-Webster.com Dictionary, Merriam-Webster, https://www.merriam-webster.com/dictionary/legacy. Accessed 2019.

"Legacy." Collins English Dictionary - Complete & Unabridged 2012 Digital Edition, https://www.dictionary.com/browse/legacy. Accessed 2019.

National Sustainable Agriculture Coalition (NSAC), *What Is the Farm Bill?* Retrieved from https://sustainableagriculture.net/our-work/campaigns/fbcampaign/what-is-the-farm-bill/.

Paisley, Clifton. (c. 1989). *The Red Hills of Florida, 1528-1865.*

Grist. Philpott, Tom. February 8, 2008. *"A reflection on the lasting legacy of 1970s USDA Secretary Earl Butz."* Retrieved from https://grist.org/article/the-butz-stops-here/.

Pierce, Charles P. (May 6, 2019). *'Rural' Is Not a Synonym for 'White,' and 'Rural Issues' Include Alleged Corporate Malfeasance and Fraud.* Esquire. Retrieved from https://www.esquire.com/news-politics/politics/a27374716/memphis-farmers-soybeans-stine-african-american/.

Potterton, Louise, IAEA Division of Public Information. International Atomic Energy Agency, News. (Sept. 16, 2011). *No Rain, No Food.* Retrieved from https://www.iaea.org/newscenter/news/no-rain-no-food/.

Quinton, Sophie. Stateline, an initiative of The Pew Charitable Trusts. (January 6, 2020) *Cannabis Confusion Pushes States to Ban Smokable Hemp.* Retrieved from https://www.pewtrusts.org/en/research-and-analysis/blogs/stateline/2020/01/06/cannabis-confusion-pushes-states-to-ban-smokable-hemp/.

Sessoms, Jasmine. Baltimore Sun. (July 9, 2019). *Lack of diversity in cannabis industry 'wholly unacceptable'.*

Tall Timbers Research Station & Land Conservancy. (c. 2020). *What is the Red Hills?* Retrieved from https://talltimbers.org/explore-the-red-hills-what-is-the-red-hills/.

"Tradition." Merriam-Webster.com Dictionary, Merriam-Webster, https://www.merriam-webster.com/dictionary/tradition. Accessed 2019.

"Tradition." Collins English Dictionary - Complete & Unabridged 2012 Digital Edition, https://www.dictionary.com/browse/tradition. Accessed 2019.

Tyler, Shakara S. and Moore, Eddie A. Professional Agricultural Workers Journal 1 (1) (2013): 6–11. *Plight of Black Farmers in the Context of USDA Farm Loan Programs.* Retrieved from https://acannabis industry companyonsearch.umn.edu/record/236726/files/Shakara%20S.%20Tyler.pdf.

U.S. Department of Agriculture, Economic Research Service. (March 2019). *The number of farms has leveled off at about 2.05 million.* Retrieved from https://www.ers.usda.gov/data-products/chart-gallery/gallery/chart-detail/?chartId=58268/.

U.S. Department of Agriculture, National Agricultural Statistics Service, Washington D.C. (April 2019). *Farm Producers.* Retrieved from https://www.nass.usda.gov/Publications/Highlights/2019/2017Census_Farm_Producers.pdf.

U.S. Department of Agriculture, National Agricultural Statistics Service. 2012 Census of Agriculture, Table 60. Selected Farm Characteristics by Race and Principal Operator: 2012 and 2007. *2012 Census of Agriculture.* Retrieved from https://www.nass.usda.gov/Publications/AgCensus/2017/Full_Report/Volume_1,_Chapter_1_US/st99_1_0060_0060.pdf.

U.S. Department of Agriculture, National Agricultural Statistics Service. 2017 Census of Agriculture, Table 61. Selected Farm Characteristics by Race: 2017. *2017 Census of Agriculture.*

205

Retrieved from
https://www.nass.usda.gov/Publications/AgCensus/2017/Full_Rep
ort/Volume_1,_Chapter_1_US/st99_1_0061_0061.pdf.

U.S. Department of Commerce, Bureau of the Census. Washington
D.C. (1933). Fifteenth Census of the United States: 1930: Census
of Agriculture: The Negro Farmer in the United States. *1930:
Census of Agriculture.* Retrieved from
https://babel.hathitrust.org/cgi/pt?id=_uiuo.ark:/13960/t9668cs41/.

Yankee Publishing, Inc. (2019). Gardening by the Moon. *The Old
Farmer's Almanac.* Retrieved from
https://www.almanac.com/content/planting-by-the-moon/.

Yeoman, Barry. (Jan. 31, 2020). *Jury awarded hog farm neighbors
$3.25 million. Will three-quarters of that be erased?* The
Charlotte Observer- Food & Environment Reporting Network.
Retrieved from https://www.
charlotteobserver.com/news/business/article239694633.html/.

Index

Editorial note: The purpose of this index is to provide the reader with a practical & concise guide to all pertinent information contained in the book. It stands as a guidepost directing the reader to a particular item in which the reader is interested. It is not meant as a glossary of every term mentioned in this book. The information in this index has been alphabetized, and in some instances, placed under groupings of main entries and subentries. These groupings correspond to the terminology and language generally used by agronomists, horticulturists, farmers, and cannabis activists when covering the topic of cannabis cultivation. Anything mentioned in passing or as background material has been omitted from this listing.

About the Author

Jimmy Jenkins is a graduate of the University of Maryland Francis King Carey School of Law, Baltimore, Maryland. He previously served as an assistant district attorney, and afterwards as a public defender. Jimmy currently works as a lawyer assisting local governments with legislative matters. He resides in the Red Hills of Leon County, Florida, with his beautiful wife, Chan, where he writes about enlightening topics based on his philosophic beliefs and life experiences.

CPSIA information can be obtained
at www.ICGtesting.com
Printed in the USA
BVHW092034151022
649514BV00003B/18